Wildland Fire Incident Management Guide 2014

January 2014

Wildland Fire Incident Management Field Guide

January 2014

PREFACE

The *Wildland Fire Incident Management Field Guide* is a revision of what used to be called the *Fireline Handbook*.

Purpose

The *Wildland Fire Incident Management Field Guide* states, references, or supplements wildland fire incident management and operational standards established by the National Wildfire Coordinating Group (NWCG).

Relationship to the Incident Response Pocket Guide *and* Emergency Responder Field Operations Guide

This field guide contains some information that is duplicated in the *Incident Response Pocket Guide* (IRPG) and the Department of Homeland Security, Federal Emergency Management Agency (FEMA), National Incident Management System *Emergency Responder Field Operating Guide* (ERFOG), but the documents have different purposes and user groups. This field guide is the standard NWCG incident management reference guide. The IRPG is the "wildland fire job aid and training reference for operational personnel from Firefighter Type 2 through Division Supervisor and Initial Attack/Extended Attack Incident Commander," and is primarily an initial responder's tool. The ERFOG "provides guidance designed to assist emergency response personnel in the use of the National Incident Management System (NIMS) Incident Command System (ICS) during incident operations" and is primarily a Command and General Staff tool.

TABLE OF CONTENTS

PREFACE 5
   Purpose 5
   Relationship to the Incident Response Pocket Guide and Emergency Responder Field Operations Guide 5
   Revision Process 5
TABLE OF CONTENTS 7
CHAPTER 1 – FIREFIGHTING SAFETY 13
   Risk Management 13
   Tenets of a High Reliability Organization 13
   Wildland Fire Safety Culture 14
   Wildland Fire Safety Principles 15
   Clothing and Personal Protective Equipment 15
   Fatigue – Work and Rest 16
   Nutrition and Hydration 16
   Driving Limitations 17
   Smoke Impairment of Roads: Assessment and Response 17
   Carbon Monoxide Poisoning 19
   Injury and Fatality Procedures 21
      Serious Injury 21
      Fatality 21
   Burn Injury Procedures 22
   Night Operations 23
   Personnel Transportation 23
   Firing Equipment 23
   Chain Saws 24
   Incident-Generated Hazmat 24
   Media Access Guidelines 25
      General Policy 25
      Guidelines 25
   Safety Responsibilities of Wildland Fire Supervisors 27
      General Responsibilities 27
CHAPTER 2 – OPERATIONAL GUIDES 29
   Initial Attack 29

Wildland Fire Incident Management Field Guide

- Definition of Initial Attack 29
- Characteristics of an Initial Attack Incident (Type 4 And Type 5 Incidents) 29
- Example of Initial Attack Organization (Type 4 Incident) 30
- Duties of an Initial Attack Incident Commander 30
- Assessing Incident Progress 32
- Updating Incident Status 32
- Fire Suppression Strategies 33
- Initial Attack Safety Checklist 34

Extended Attack 35
- Definition of Extended Attack 35
- Characteristics of an Extended Attack Incident 35
- Example of an Extended Attack Organization 36
- Change From an Initial Attack Incident to an Extended Attack Incident 36
- Control or Transfer to Type 2 Incident 38
- Extended Attack Safety Checklist 39

Large Fire Management Teams 40
- Type 2 Organization 40
- Type 1 Organization 40
- Organization Chart for Type 1 and Type 2 Incidents 41
- Area Command 41
- Unified Command 41

Transfer of Command 42
- Incident Commander Briefing 42
- Incident Commander's Checklist 43
- Agency Administrator(s)' Responsibility for the Transfer of Command and Release of Incident Management Teams 43
- Transfer of Authority 44
- Agency Administrator Briefing 44
- Release of an Incident Management Team 45

Urban Interface 46
- Wildland/Urban Interface "Watch Out" Situations 46
- Identification of Reduced-Risk Structures and Communities 46
- Structure Triage Guidelines 47
- Structure Assessment Checklist (if Time Permits) 48
- Structure Protection Guidelines 50

CHAPTER 3 – POSITION RESPONSIBILITIES 53
Command and General Staff 53
- Organization Chart 53
- Position Checklists 54

Operations 58
- Organization Chart 58
- Position Checklists 59
- Air Operations 64
- Position Checklists 66
- Helispot Location and Construction 74
- Principles of Retardant Application 75

Planning 76
- Organization Chart 76
- Position Checklists 77
- Planning Process 89
- Demobilization 89

Logistics 90
- Organization Chart 90
- Position Checklists 91
- Logistics Guidelines 99
- Factors to Consider When Locating and Laying Out an Incident Base or Camp 101

Finance/Administration 102
- Organization Chart 102
- Position Checklists 103

CHAPTER 4 – REFERENCE 109
Portable Pumps and Hydraulics 109
- Formula for Determining Pump Pressure 109
- Reminders for Using Portable Pumps and Hose Lays 110
- Drafting Guidelines 110
- Expected Output of Commonly Used Portable Pumps at Sea Level 110
- General Rules for Fireline Hydraulics 111
- Friction Loss by Hose Size and Type 112
- Pump Pressures for 50-psi Nozzle Pressure 113
- Pump Pressures for 50-psi Nozzle Pressure 114

- Pump Pressures for 50-psi Nozzle Pressure (Continued) 115
- Foam 116
  - Foam Use 116
  - Foam Mixture Rates 116
  - Foam for Direct Attack 117
  - Foam for Indirect Attack 117
  - Foam for Mop Up 117
  - Foam for Exposure Protection 117
  - Foam Safety 118
- Use of Fireline Explosives 118
  - Advantages 118
  - Disadvantages 118
- Hazmat Checklist for Incident Base Management 119
- Use of Inmate Crews 120
- Production Tables 121
  - Sustained Line Production Rates of 20-Person Crews in Feet per Hour* 121
  - Sustained Line Production Rates of 20-Person Crews in Feet per Hour* 121
  - Sustained Line Production Rates of 20-Person Crews in Chains per Hour* 122
  - Sustained Line Production Rates of 20-Person Crews in Chains per Hour* 122
  - Line Production Rates for Initial Action by Hand Crews in Chains per Person per Hour 123
  - Line Production Rates for Initial Action by Engine Crews in Chains per Crew per Hour 124
  - Fireline Explosives Production Comparisons 125
  - Dozer Fireline Construction Rates (Single Pass) in Chains per Hour 126
  - Dozer Fireline Construction Rates (Single Pass) in Chains Per Hour (Continued) 127
  - Tractor Plow Fireline Production Rates in Chains per Hour 128
- Interagency Crew Qualifications and Equipment Standards 129
  - Minimum Crew Standards for National Mobilization 129
  - NWCG Engine and Water Tender Typing (Minimum Requirements) 130
  - Common Additional Needs for Engines and Tenders (Request As Needed) 131
  - Air Tankers 132
  - Helicopters 132
- Other References 133
  - Clear-Text Guide 133
  - Clear-Text Guide (Continued) 134

- ICS Map Display Symbols 135
- ICS Map Display Symbols (Continued) 136
- Conversion Factors for Map Scale 138
- Formula for Area and Circumference of a Circle 138
- Acreage Determination Factors 139
- Conversion Factors 142
- Conversion Factors (Continued) 143
- Incident Command System Forms 144
- Resource Status Card 145
- Distances and Formulas for Estimating Fire Size 146
- What the Color and Column of Smoke May Mean 147
- Fire Suppression Interpretations from Flame Length 148
- Wildland Fire Risk and Complexity Assessment 149
- Indicators of Incident Complexity 153

ACRONYMS 155

This page intentionally left blank.

CHAPTER 1 – FIREFIGHTING SAFETY

Firefighter and public safety is the first priority of the wildland fire management program and must always take precedence over property and resource loss.

RISK MANAGEMENT

The wildland fire environment possesses inherent hazards that can result in harm to firefighters engaged in fire operations. Therefore, sound risk management is the foundation for all fire management activities. Risk management is defined as the process whereby management decisions are made and actions taken concerning the control of hazards and acceptance of remaining risk. The risks involved with any fire activity must be identified, assessed, and mitigated (or eliminated) when possible and practicable. The remaining risk must be considered acceptable to everyone involved and be weighed against the potential benefit during the management decision of continuing or discontinuing the activity. We practice risk management to minimize firefighters' exposure to inherent hazards in fire operations while still accomplishing management objectives.

The five-step risk management process is outlined in the *IRPG*.

Step 1 – Establish situation awareness.
Step 2 – Identify hazards and benefits and assess the risk.
Step 3 – Control, mitigate, or eliminate hazard.
Step 4 – Make go/no-go decision based on acceptability of remaining risk.
Step 5 – Evaluate effectiveness of hazard controls and continuously reevaluate.

TENETS OF A HIGH RELIABILITY ORGANIZATION

Wildland firefighting is described as a "high reliability organization" (HRO) in the book *Managing the Unexpected*, by Karl Weick and Kathleen Sutcliffe. The term "high reliability organization" came from researchers who studied the operations and culture of organizations that routinely operate in high risk environments yet endure less than their fair share of accidents.

The tenets of an HRO serve as foundational practices that firefighters at all levels should promote and model. They include:

- **Preoccupation With Failure** – Identify and report all close calls, near misses, and unexpected outcomes. They may be symptomatic of a problem with the system and could lead to a serious accident if several separate small errors coincide.

- **Reluctance to Simplify** – Invite skepticism to common practices, question standards and procedures, and reconcile diverse opinions.

- **Sensitivity to Operations** – Be vigilant to the early detection of small errors, especially on the fireline. SPEAK UP if you see potential hazards or anything that seems unexpected. Make continuous adjustments to operations that prevent errors from accumulating.

- **Commitment to Resilience** – Think ahead and ask "what if" questions. Always consider the worst-case scenario. Identify seemingly small errors and find solutions to unexpected problems. Maintain the capacity for dealing with new or changing circumstances.

- **Deference to Expertise** – Allow decisions to be made by those with the most expertise, not necessarily the one with the most experience or highest rank. Promote and cultivate diversity of thought, and express your opinion.

WILDLAND FIRE SAFETY CULTURE

The wildland fire safety culture promotes the concept of organizational learning. Firefighters at all levels should view close calls, near-misses, and/or unintended outcomes as an opportunity to collectively better our understanding of the systematic conditions in which we operate. By openly reporting and examining these events, we gain valuable knowledge of previously unrecognized risks and conditions that lend themselves to potential risks before a serious accident occurs. The tenets of an HRO are predicated on an organizational "safety culture" that strives to identify and minimize previously unrecognized risk through collective organizational learning. The four critical subcomponents of a safety culture outlined in James Reason's *Managing the Risks of Organizational Accidents* include:

- **Reporting Culture** – Firefighters at all levels should report, openly or anonymously, all critical or near miss incidents through SAFENETs, SAFECOMs, or Activity Logs (ICS 214) even if they entail reporting firefighters' own errors. Supervisors should, as far as practicable, separate the collection of information and analysis of reports for safety purposes from any law enforcement investigation or disciplinary proceedings that may follow a serious accident.

- **Just Culture** – Fire Management should promote an atmosphere of trust where people are encouraged to provide essential safety-related information, yet draw a clear line between acceptable and unacceptable behavior.

- **Flexible Culture** – Firefighters at all levels are encouraged to adapt to changing conditions and are given the latitude to exercise judgment. They should be technically proficient in their job but never assume they are prepared for all contingencies. Expect the unexpected and react accordingly.

- **Learning Culture** – Firefighters should seek continuous, collective improvement in wildland fire safety by observing conditions at all levels of the organization that may lead to potential risks. They should report and help create safety mitigations that can prevent future occurrences.

WILDLAND FIRE SAFETY PRINCIPLES

Following a systematic risk management process, along with adherence to common safety principles and the willingness to identify and report potentially hazardous conditions, are the foremost responsibilities of every firefighter, supervisor, and administrator at all operational levels.

Most of the common fire safety principles for wildland fire operations are found in the *Incident Response Pocket Guide* (*IRPG*) which should be carried by all operational firefighters on the fireline. These safety principles should be understood at all levels of command. They include but are not limited to:

- Risk Management Process
- Look Up, Look Down, Look Around
- Standard Firefighting Orders
- Watch Out Situations
- Lookout(s), Communication(s), Escape(s), and Safety Zone(s) (LCES)
- Safety Zone Guidelines
- Downhill Line Construction Checklist
- Communication Responsibilities
- First Aid Guidelines

CLOTHING AND PERSONAL PROTECTIVE EQUIPMENT

- All personal protective equipment (PPE) will meet or exceed agency policy.
- Wear flame-resistant clothing on the fireline and when flying in helicopters. Do not wear clothing, even undergarments, made of synthetic materials, which can burn and melt on your skin. Roll your sleeves down to the wrist.
- Flame-resistant clothing should be cleaned or replaced whenever soiled, especially when soiled with petroleum products. Flame-resistant clothing will be replaced when the fabric is so worn as to reduce the protection capability of the garment, or is so faded as to significantly reduce the desired visibility qualities. Yellow long-sleeved aramid shirts are required for national mobilization.
- Wear a hard hat and leather gloves while on the fireline.
- Personnel assigned to wildland fires must wear a minimum of 8-inch-high, laced-type exterior work boots, with Vibram-type, melt-resistant soles. The 8-inch height requirement is measured from the bottom of the heel to the top of the boot. Alaska is exempt from the Vibram-type sole requirement.
- Use eye and face protection whenever there is a danger from material being thrown back in your face.

Chapter 1 – Firefighting Safety

- Determine and comply with host agency requirements regarding fire shelters on fireline suppression assignments or follow your own agency's requirements if they are more restrictive.

- Use hearing protection when working with high-noise-level firefighting equipment, such as helicopters, air tankers, chain saws, pumps, etc.

- When operating chain saws, sawyers and swampers will wear additional safety equipment, including approved chaps, gloves, hard hat, and eye and hearing protection. Swampers should wear chaps when the need is demonstrated by a risk analysis considering proximity of the chain saw to the sawyer, and the slope, fuel type, etc.

- Face and neck protection (Nomex shrouds) are not required PPE. If issued, shrouds should be deployed only in impending flash fuel or high-radiant heat situations and not routinely worn throughout the operational period, due to an unacceptable increase in physiological heat stress.

FATIGUE – WORK AND REST

- For adequate sleep and rest environment, plan to provide 1 hour of sleep or rest for every 2 hours worked. Monitor individuals for elevated levels of fatigue.

- When deviating from work/rest guidelines, the Agency Administrator or Incident Commander (IC) must approve in writing.

NUTRITION AND HYDRATION

- A firefighter may burn 5,000 to 6,000 calories a day. These calories must be replaced to help avoid fatigue and impaired judgment.

- Government-provided food should be primarily carbohydrates. The recommended ratio of macronutrients is 60% carbohydrates, 10% protein, and 30% fat (less than 10% should be saturated fats). Only macronutrients provide energy. During fireline assignments, firefighters should try to eat 160 calories of carbohydrates per hour.

- On a normal fireline assignment, firefighters may need to replace 5-6 quarts of fluids a day. Mix water with natural juices and sport drinks containing energy-restoring glucose to help meet the 160 calories of carbohydrate per hour recommendation. It is important to remember that each individual is different, and there are no recommendations that apply to everyone exactly.

DRIVING LIMITATIONS

Drivers operating vehicles that require a Commercial Driver's License (CDL) are regulated by the Federal Motor Carriers Safety Regulations, Part 393.3, and any applicable state laws. Operators must adhere to guidelines in NWCG memo dated February 6, 2004, or Department of Transportation (DOT) regulations, whichever is more restrictive.

The February 6, 2004, NWCG memo on Interagency Driving Standards states: "These standards address driving by personnel actively engaged in wildland fire or all-risk response activities, including driving while assigned to a specific incident or during Initial Attack fire response (includes time required to control the fire and travel to a rest location). In the absence of more restrictive agency policy, these guidelines will be followed during mobilization and demobilization as well. Individual agency driving policies shall be consulted for all other non-incident driving.

1. Agency resources assigned to an incident or engaged in Initial Attack fire response will adhere to the current agency work/rest policy for determining length of duty day.
2. No driver will drive more than 10 hours (behind the wheel) within any duty-day.
3. Multiple drivers in a single vehicle may drive up to the duty-day limitation provided no driver exceeds the individual driving (behind the wheel) time limitation of 10 hours.
4. A driver shall drive only if they have had at least 8 consecutive hours off duty before beginning a shift. Exception: to the minimum off-duty hour requirement is allowed when **essential** to:
    a. accomplish immediate and critical suppression objectives, or
    b. address **immediate** and **critical** firefighter or public safety issues.
5. As stated in the current agency work/rest policy, documentation of mitigation measures used to reduce fatigue is required for drivers who exceed 16-hour work shifts. This is required regardless of whether the driver was still compliant with the 10-hour individual (behind the wheel) driving time limitations."

SMOKE IMPAIRMENT OF ROADS: ASSESSMENT AND RESPONSE

Smoke has the potential to cause severe safety hazards to vehicle traffic in the vicinity of active fires, especially at night. Minimum Acceptable Vehicle (MAV) is the distance that includes the necessary braking (stopping) distance required for the posted speed limit plus the normal reaction distance that is covered before brakes are actually applied. When smoke from wildfires impacts roads, the MAV should be determined and an immediate evaluation of environmental conditions for fog formation should be completed. Also, the hours when these conditions simultaneously occur need to be noted. In general, weather affects how smoke disperses on roads in the following ways:

- Low surface wind speed (<7 mph or calm or light winds)
- Lack of cloud cover (<40%; clear skies facilitate rapid cooling)

- Rapid cooling at the ground surface (2 to 3 degrees per hour for 3 to 6 hours after sunset or temperature is <70 °F)
- Relative Humidity (RH)
  - ✓ >70% in the presence of smoke will begin to severely reduce visibility,
  - ✓ >80% smoke-induced fog formation can be expected, and
  - ✓ >90% natural fog formation can be expected.

Even when smoke does not reach extreme levels, visibility on roads may be reduced and may still be a safety hazard that requires a response and mitigation to protect personnel and the public.

- When potential smoke-related problems are identified:
  - ✓ Advise the Agency Administrator that severe smoke conditions exist.
  - ✓ Implement preplanned actions, such as posting "smoke warning" signs.
  - ✓ Ensure proper equipment is ready and appropriate personnel are briefed on contingency plans and are available to control traffic.
  - ✓ Notify local law enforcement units of the potential problem.
- Establish periodic patrols to monitor smoke-impacted areas.
- When smoke-related traffic problems occur, the first person on the scene must maintain traffic control until relieved by someone else or the smoke or traffic problem no longer exists. He or she should take immediate action, such as the following, to prevent injuries and damage:
  - ✓ Establish control points on both sides of the impacted area.
  - ✓ Slow or stop traffic entering the area. Advise drivers of alternate routes.
  - ✓ Assign a person to keep a log of what actions are taken.
  - ✓ Ensure warning signs are in place and any other preplanned actions have been implemented.
  - ✓ Notify personnel identified and equipped to direct traffic. Notify other local units having responsibilities for traffic control.
  - ✓ Implement radio and television traffic advisories for the impacted area.
- Smoke moving unexpectedly into an area may be an indication of changing burning conditions. All traffic should be excluded until this change can be evaluated.
- When smoke-related traffic accidents occur, fire personnel on the scene should:
  - ✓ Make all efforts to assist and protect people.
  - ✓ Notify, if necessary, appropriate medical units and request assistance.
  - ✓ Notify appropriate law enforcement units.
  - ✓ Provide additional personnel for traffic control, if necessary.

- ✓ Notify Agency Administrator who may assign local safety and tort claims personnel to the scene.
- Assign an individual (preferably a law enforcement official) to record facts about the accident, including names, addresses, and statements of witnesses (if given willingly). At a minimum, record license plate identification on all vehicles in the vicinity of the accident. Coordinate efforts with local law enforcement personnel.
  - ✓ Fire personnel at accident scene, if questioned by someone other than law enforcement officers, should only state that their involvement was in fire suppression activities in the vicinity.
- Involved personnel should submit written reports of their actions and observations immediately after being released from the accident scene.

CARBON MONOXIDE POISONING

Carbon monoxide (CO) is an odorless, tasteless, invisible gas by-product emitted from combustion of forest and range fuels, internal combustion engines, and a variety of other sources. In a wildfire, elevated concentrations of CO can coexist with smoke. Elevated concentrations of CO have been associated with the following operational tasks: mop up, direct line construction, and holding operations. Common symptoms of CO exposure are headache, nausea, rapid breathing, weakness, exhaustion, dizziness, and confusion. If a firefighter experiences these symptoms, the firefighter should notify his or her supervisor.

To manage CO exposure:

- Monitor workers, particularly pump and chain saw operators, for symptoms or behavior associated with CO exposure.

Blood CO Level	Symptom	Behavior
Moderate	Possible headache, nausea, and increasing fatigue.	Increasing impairment of alertness, vision discrimination, judgment of time, and physical coordination. Becomes increasingly complacent.
High	Headache, fatigue, drowsiness, nausea, vomiting, dizziness, convulsions, cardiorespiratory difficulty.	Above behavior becomes more acute to extreme.

- Remove workers from work site to "CO-free areas" when performance and safety are compromised by symptoms or behavior described above.
- When possible, select strategy and tactics that minimize worker exposure to smoke concentrations (indirect attack). Expect higher CO concentrations in the following:
  - ✓ Near an active flame front and during smoldering phase of combustion.
  - ✓ Working around heavy equipment, especially in ground support.

- ✓ Heavy smoke concentrations during inversions or areas downwind of the fire. Mop up (prolonged exposure to low to moderate smoke level) can increase risk of overexposure to CO and particulate matter.
- ✓ Topographic features that concentrate smoke (head of canyon, ravines, saddles or passes, depressions or basins).

- Periodically rotate workers from work sites with moderate to high smoke levels to areas of less smoke or smoke-free areas.
- If necessary, order additional personnel to relieve crews assigned to high-smoke-level areas.
- Instruct personnel to take breaks in smoke-free or low-smoke areas, when possible.
- Locate incident Base and Camp(s) in areas free of smoke and air pollution to maximize recovery from CO exposure.
- Portable propane heaters are not to be used inside tents, yurts, or other enclosed spaces.
- Encourage smokers to terminate or reduce smoking during fire assignment. Smoking significantly increases blood CO levels.
- Restrict workers from driving a vehicle if they display the symptoms or behavior outlined above.
- Personnel who display the symptoms or behavior outlined above should be evaluated and determined fit for duty before their next work assignment.

INJURY AND FATALITY PROCEDURES

Serious Injury

- Give first aid. Call for medical aid and transportation if needed.
- Do not release victim's name except to authorities.
- Never broadcast victim's name on the radio.
- Do not allow unauthorized picture taking or release of pictures.
- Notify IC, who will:
  - ✓ Assign a person to supervise evacuation, if necessary, and stay with the victim until under medical care. In rough terrain, at least 15 workers will be required to carry a stretcher.
  - ✓ Assign a person to get the facts and witness statements and preserve evidence until investigation can be taken over by the Safety Officer (SOFR) or appointed investigating team.
  - ✓ Notify the Agency Administrator.

Fatality

- Do not move the body unless it is in a location where it could be burned or otherwise destroyed. Secure the accident scene.
- Do not release victim's name except to authorities.
- Never broadcast victim's name on air.
- Do not allow unauthorized picture taking or release of pictures.
- Notify the IC, who will:
  - ✓ Assign a person to start the investigation until relieved by an appointed investigating team.
  - ✓ Notify the Agency Administrator and report essential facts. The Agency Administrator will notify proper authorities and next of kin as prescribed by agency regulations.
  - ✓ If requested, assist authorities in transporting remains. Mark location of the body on the ground. Note location of tools, equipment, or personal gear.
  - ✓ Retain PPE as evidence.

Chapter 1 – Firefighting Safety

BURN INJURY PROCEDURES

After onsite medical response, initial medical stabilization, and evaluation are completed, the Agency Administrator or designee having jurisdiction for the incident and/or firefighter representative (e.g., Crew Boss, Medical Unit Leader, Compensation-For-Injury Specialist, etc.) should coordinate with the attending physician to ensure that a firefighter whose injuries meet any of the following burn injury criteria is immediately referred to the nearest regional burn center. It is imperative that action is expeditious, as burn injuries are often difficult to evaluate and may take 72 hours to manifest themselves. These criteria are based upon American Burn Association criteria as warranting immediate referral to an accredited burn center.

The decision to refer the firefighter to a regional burn center is made directly by the attending physician or may be requested of the physician by the Agency Administrator or designee having jurisdiction and/or firefighter representative.

The Agency Administrator or designee for the incident will coordinate with the employee's home unit to identify a Workers Compensation liaison to assist the injured employee with workers compensation claims and procedures.

Workers Compensation benefits may be denied in the event that the attending physician **does not agree** to refer the firefighter to a regional burn center. During these rare events, close consultation must occur between the attending physician, the firefighter, the Agency Administrator or designee, or firefighter representative, and the firefighter's physician to assure that the best possible care for the burn injuries is provided.

Burn injury criteria:

Partial thickness burns (second degree) involving more than 5% Total Body Surface Area (TBSA).

- Burns (second degree) involving the face, hands, feet, genitalia, perineum, or major joints.
- Third-degree burns of any size are present.
- Electrical burns, including lightning injury, are present.
- Inhalation injury is suspected.
- Burns are accompanied by traumatic injury (such as fractures).
- Individuals are unable to immediately return to full duty.

When there is any doubt as to the severity of the burn injury, the recommended action should be to facilitate the immediate referral and transport of the firefighter to the nearest burn center.

A list of possible burn care facilities can be found on the NIFC Web site at:
http://www.blm.gov/nifc/st/en/prog/fire/im.html

For additional incident emergency medical information see the web site at:
http://www.nwcg.gov/branches/pre/rmc/iems/index.html

NIGHT OPERATIONS

Every effort shall be made to orient work crews scheduled for night operations during daylight hours and provide adequate lights and communication. A knowledgeable day operations representative should remain on site to properly orient and brief night operations crews, particularly about line location and boundaries, predicted weather, fire behavior, terrain features, hazards, and control problem areas.

PERSONNEL TRANSPORTATION

- Overhead should have a driver whenever possible.
- All passengers in vehicles must be seated and seatbelted with arms and legs inside vehicle.
- Personnel and unsecured tools will not be transported together.
- Driver must be qualified for the vehicle and operating conditions. If not, remove them from driving duties.
- When traveling to a fire, observe all traffic signals, speed limits, and safety rules.
- Driver should walk around vehicle to make sure all is clear before departure.
- Driver is responsible for arrangements to ensure that if chock blocks are provided, they are in place before loading, unloading, or when parked.
- When transporting personnel, the driver shall not leave his or her seat until the vehicle is securely chocked. **NEVER** load or unload personnel from an **UNCHOCKED VEHICLE**.
- Driver shall conduct a daily mechanical check of vehicle before driving. Unsafe equipment should be removed from service and reported to the Ground Support Unit for repair.
- Driver should use spotter outside of vehicle when backing or turning around.
- Driver should operate vehicles with headlights on at all times.

FIRING EQUIPMENT

- Only trained personnel should use firing equipment.
- Use only approved equipment and qualified personnel when firing from helicopters.
- Use no more than one part gasoline to three parts diesel (or heavier fuel) in flamethrower or drip torches. Observe manufacturers' recommendations.
- When operating ground-based firing equipment that uses jellied gasoline, to avoid back splatter, do not direct the stream of burning material into the tops of nearby trees or tall brush.
- Properly ground firing equipment when fueling.
- Maintain constant radio communications between the firing operation and other appropriate fireline personnel.

CHAIN SAWS

- Stop engine when carrying, making adjustments, repairing, or cleaning a chain saw.
- Use bar guards when carrying saw in rough country.
- Cool chain saw engine before refueling. Fill on bare ground and move at least 10 feet from fueling area before starting.
- Use proper safety equipment, such as chaps, gloves, hard hat, and eye and hearing protection. Swampers should wear chaps when the need is demonstrated by a risk analysis considering proximity to the sawyer, slope, fuel type, etc.

INCIDENT-GENERATED HAZMAT

Firefighters, supervisors, and agency representatives are not necessarily aware of the dangers of transporting hazardous materials. Many of these materials, used frequently on the fire job, are not considered hazardous by firefighters.

Petroleum products, especially gasoline, are prohibited from public-transportation vehicles because of the obvious danger. Crews should not transport petroleum products inside enclosed vehicles. Gasoline should be purged from all gas cans, chain saws, etc., before transport.

Other items, such as ignition devices, fusees, explosives, and mineral spirits, cannot be placed on commercial aircraft or other public transportation.

Supply and Ground Support Unit Leaders should be well trained in handling hazardous materials and should make provisions at the incident to ensure petroleum containers are purged and fusees are left at the incident for safe return to the cache.

Supply and Ground Support Unit Leaders should be made aware of standard transportation rules regarding hazardous materials. For instance, oxidants, such as fertilizer, should not be transported with flammables. Be careful not to mix incompatible materials (for example, ammonia should not be transported with chlorine). All packages must be clearly labeled according to Department of Transportation (DOT) and Occupational Safety and Health Administration (OSHA) standards. All packages and containers should also be checked thoroughly for damage and leaks. Some spills can be more dangerous than expected.

Incident needs may require transportation of hazardous materials from Base or Camp to the fireline. Basic knowledge of how to safely handle a variety of flammables, oxidants, cleaners, etc., should be taught to all fire personnel.

MEDIA ACCESS GUIDELINES

General Policy

- It is the policy of Federal and state agencies to provide news media access to incidents, including wildland fires and prescribed fires.
- Federal and state agencies are required to provide equitable and maximum news media access to wildland fire incidents.
- For the purposes of these guidelines, news media representatives include print and broadcast reporters, freelance print reporters, freelance videographers, and photographers.
- While the wildland firefighting agencies seek to provide safe access to incidents for news media representatives, the ultimate responsibility for their safety lies with the individual reporter and their employer.

Guidelines

Access

- Visits to the fireline must receive the approval of the IC or designated representative.
- News media representatives will be escorted by a person qualified as a Single Resource Boss or other appropriate escort approved by the IC. The IC may delegate escort approval authority to other incident personnel, such as the lead Public Information Officer (PIOF) or appropriate local authority.

Personal Protective Equipment

- News media representatives will be required to wear PPE as outlined in this field guide and in the *Interagency Standards for Fire and Fire Aviation Operations* (the "Red Book") when working on or near the fireline, and have an appropriate safety briefing. PPE must meet National Fire Protection Association (NFPA) and National Wildfire Coordinating Group (NWCG) standards. The required PPE is:
    - ✓ 8-inch-high, lace-type work boots with nonslip, melt-resistant soles and heels
    - ✓ Aramid shirts
    - ✓ Aramid trousers
    - ✓ Hard hat with chinstrap
    - ✓ Leather gloves
    - ✓ Fire shelter
    - ✓ Water canteen

    PPE may be provided by the fire organization if media representatives are unprepared.

Firefighter Training

- News media representatives are invited to join in basic firefighter courses with Federal and state providers, if there is sufficient room for them. These courses are voluntary. News media should be informed that attending the courses is not a guarantee of access or an endorsement of safety accreditation, but rather is an opportunity for information and education. Reporters can also be referred to authorized contract trainers or the academic community for basic firefighter courses.
- NWCG member agencies will not administer the work capacity test to news media representatives because of liability concerns.

"Shadowing" Fire Crews

- Personnel assigned to an incident will facilitate indepth coverage opportunities for journalists. News media representatives requesting to "shadow" crews for more than one operational period on the fireline or in the fire area must:
  - ✓ Wear PPE and understand how to use it in accordance with the direction in this field guide.
  - ✓ Coordinate activities with the lead PIOF, who will communicate with the affected Crew Boss, IC, and the Fire Management Officer at the crew's home unit.
- It is strongly recommended that reporters requesting to shadow crews complete basic firefighter training, including S-130 and S-190. If these courses have been taken in a previous year, a current refresher course is recommended. News media representatives must be able to affirm that they can walk in mountainous terrain, are in good physical condition, and have no known physical limitations.

Red Cards in the Incident Command System

- News media representatives will not be issued Incident Qualification Cards or "red cards" under the Incident Command System. The red-card system was designed for incident personnel with specific duties for which they are trained and qualified, and not for personnel not officially assigned to the incident.

Existing Laws and Policies

- These guidelines apply to all wildland fires and prescribed fires under Federal or state jurisdiction, but are not intended to supersede existing tribal laws; state laws, such as media access laws in California; or chain-of-command procedures applicant to military crews.

Denial of Access

- Denial of access to fire camp, the fireline, or other related areas will be a rare occurrence. News media access may be limited when the IC determines:
  - ✓ Safety of firefighters or others may be compromised. Considerations should be the same as those for determining that conditions are unsafe for fire crews to be on the fireline, including extreme fire behavior or expected change in the weather.
  - ✓ The presence of nonfire personnel compromises incident operations.
  - ✓ The presence of nonfire personnel compromises the integrity of an investigation.
  - ✓ A violation of security or privacy of incident personnel would occur.
- Federal and state agencies will support decisions regarding access by other jurisdictions, such as a private landowner, tribal entity, or local law enforcement agency, such as when the local law enforcement agency closes an area for evacuation purposes.
  Reasons for denial of access should be documented by the lead PIOF and become part of the Unit Log.
- News media aviation resources must determine and abide by airspace restrictions that may be implemented by the Federal Aviation Administration (FAA) at the request of fire managers.

SAFETY RESPONSIBILITIES OF WILDLAND FIRE SUPERVISORS

General Responsibilities

Personal actions describe safety more effectively than written plans or "rule books." Firefighters' actions tell what they consider important. Model good safety habits and demand the same from your subordinates.

Supervision of other firefighters includes the following tasks:

- Maintain accountability of assigned personnel's exact location and general welfare at all times, especially during incident operations.
- Set a personal example of safe behavior and enforce safe practices.
- Assign fireline assignments only to people who are properly qualified and physically fit for the job.
- Evaluate firefighters' physical and mental condition.
- Analyze work situations to eliminate or avoid hazards. Discuss safety at the beginning of each shift or new work assignment.
- Become immediately involved whenever an injury occurs, and ensure that medical treatment is provided in a timely manner.
- Monitor work to be sure it is done safely and efficiently.
- Monitor and enforce work/rest guidelines.

- Provide leadership in applying corrective action aimed at eliminating accidents and instilling a safe work attitude.
- Protect employees from reprisal for reporting unsafe conditions.

REMEMBER: EACH INDIVIDUAL, AND ESPECIALLY SUPERVISORS, HAVE AND MUST RECOGNIZE THEIR SAFETY RESPONSIBILITIES.

CHAPTER 2 – OPERATIONAL GUIDES

INITIAL ATTACK

Definition of Initial Attack

Initial Attack is the actions taken by the first resources to arrive at a wildfire or wildland fire use incident. Initial actions may be size up, patrolling, monitoring, holding action, or aggressive Initial Attack. All wildland fires that are controlled by suppression forces undergo Initial Attack. The kind and number of resources responding to Initial Attack vary depending upon fire danger, fuel type, values to be protected, and other factors. Generally, Initial Attack involves a small number of resources, and incident size is small. **REGARDLESS OF FIRE TYPE, LOCATION, OR PROPERTY OR RESOURCE, BEING THREATENED, FIREFIGHTER SAFETY WILL ALWAYS BE THE #1 PRIORITY.**

Characteristics of an Initial Attack Incident
(Type 4 And Type 5 Incidents)

- Resources vary from a single resource (Type 5) to several single resources (Type 4), possibly a single Strike Team or Task Force.

- Normally limited to one operational period, at least the containment phase. Mop up and/or control may extend into multiple periods.

- Normally does not require a written Incident Action Plan (IAP). May use the ICS Incident Briefing Form (ICS 201).

- The Initial Attack Incident Commander (ICT4 and ICT5) is responsible for performing all Command and General Staff functions.

Example of Initial Attack Organization (Type 4 Incident)

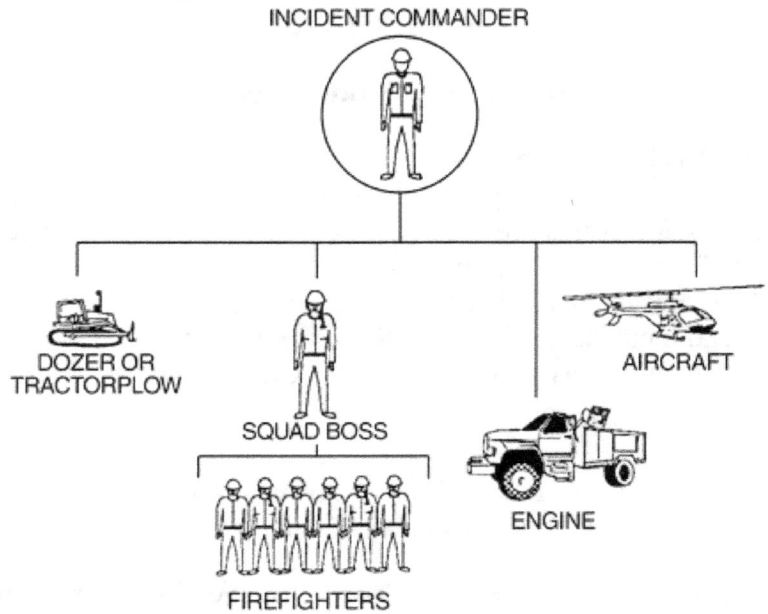

Duties of an Initial Attack Incident Commander

Upon Dispatch

Obtain as much of the following incident information as possible when dispatched to a wildland fire:
- Fire location
- Other jurisdiction(s) involved
- Best access
- Size
- Values threatened
- Person reporting the fire
- Fuels involved
- Hazards
- Current fire weather information, based on latest fire weather forecast
- Landowner, if available
- Fire cause, if available
- Rate of spread

En Route to Incident
- Travel safely! Do not speed!
- Consider what you know about the area.
- Consider expected fire behavior.
- Consider whether observed fire weather matches forecasts.
- Consider whether to Request a Spot Weather Forecast.

Arriving On Scene

When Approaching the Scene
- Use caution when approaching the scene. Observe fire scene for "Look Up, Look Down, Look Around" concerns.
- Watch for people leaving the area; take information that may assist with a fire investigation.
- Identify best access routes into fire and escape routes; pass information on to incoming resources.

Once On Scene
- Establish command of the fire.
- Establish a command organization, e.g., ICS.
- Advise Dispatch and onscene resources that you are on scene and assuming command.
- Provide Dispatch with a sizeup report
- Initiate risk management process. (Refer to the *IRPG*).

DO NOT CROSS THE FIRE'S HEAD UNLESS IT CAN BE DONE SAFELY!
- Ensure that access into the fire scene is kept open.
- Attempt to locate fire origin, and protect it.
- Account for all personnel and equipment that are already on scene.
- Review Initial Attack Safety Checklist.
- Using the information from the fire sizeup, develop incident objectives and fire suppression strategies, and ensure that assigned personnel know them.
- Provide an initial briefing of resources at or arriving on scene using the briefing checklist from the inside back cover of the *IRPG*.
- Make sure personnel understand their assignment before going to work.
- Ensure that all responders are wearing the appropriate PPE.

FIRES SHOULD BE FOUGHT AGGRESSIVELY, BUT SAFETY AND PROTECTION OF PERSONNEL AND EQUIPMENT MUST BE THE TOP PRIORITY.

REMEMBER: STANDARD FIREFIGHTING ORDERS, LCES, AND WATCH OUT SITUATIONS

Assessing Incident Progress

- After resources have been deployed and suppression actions started, assess incident progress and make any necessary changes to the IAP.
- Make sure all affected resources are advised of IAP changes.
- Assess whether incident size and complexity have reached a level where you are no longer qualified as an IC. Review the Initial Attack Safety Checklist as needed or when incident conditions change.

DO NOT HESITATE TO ASK FOR HELP!

Updating Incident Status

At the earliest opportunity, forward the following incident information to the agency Dispatch (continue to keep Dispatch updated of any significant changes and progress on the fire):

- Actual location
- Size of fire
- Rate of spread
- Fire potential (how large will or may the fire get?)
- Anticipated control problems
- Estimated control time
- Values threatened
- Fuel type
- Topography
- Weather conditions (especially if different from initial report)
- Resources on scene
- Additional resource needs
- Resource released
- Cause (if known)

Fire Suppression Strategies

The strategy(s) used to control a fire depend on the rate of spread, intensity, spotting potential, values at risk, size, type of available resources, and other factors. Anchor control lines to an existing barrier such as a road, creek, burned area, etc., to minimize the chance of being flanked by the fire. Suppression action(s) may include one or a combination of the following strategies:

Direct Attack

- Used when fire perimeter is burning at low intensity and fuels are light, allowing for safe operation at the fire's edge.
- Control efforts, including line construction, are done at the fire perimeter, which becomes the control line.
- Unless special situations dictate otherwise, line construction will start from an anchor point. **KEEP ONE FOOT IN THE BLACK WHEN POSSIBLE.**

Indirect Attack

- Used when a direct attack is not possible or practical.
- Fireline is located some distance from fire's edge.
- Terrain, fuels, fire behavior, and available resources will dictate fireline placement.
- Burning out of indirect line is handled as a second phase of line construction.

Initial Attack Safety Checklist

Answer the following questions (repeat this checklist whenever there is a change in conditions on the fire or a predicted change in fire conditions). If the answer is NO to any of the checklist questions, you MUST take the appropriate corrective action(s) IMMEDIATELY.

Yes	No	Question
		Does everyone (Dispatch and onscene resources) know who the Incident Commander is?
		Have you sized up the fire and established Incident Objectives?
		Have you initiated the Risk Management Process? (See chapter 1, page 1)
		Do you have a current fire weather forecast for fire location?
		Is the observed fire weather consistent with the forecast?
		Have you developed a plan to attack the fire (direct or indirect, anchor points, priority areas)? Have you communicated this plan to all personnel assigned to the fire, including new arrivals?
		Can you control the fire with the resources available (onscene and en route) under expected conditions?
		Do you have a sufficient command organization in place?
		Do you have a complete list of onscene and incoming resources?
		Are Watch Out Situations and Standard Firefighting Orders being followed?
		Are Lookouts in place or you can see all of the fire area?
		Can you Communicate with everyone on the fire and with Dispatch?
		Are adequate Escape routes and Safety zones established?
		Will you control the fire before the next operational period?
		Have you reported the fire's status to Dispatch?
		If the fire will not be controlled before the next operational period, have you informed agency headquarters?
		Do the fire size or complexities remain within your capabilities and qualifications to manage the fire?

EXTENDED ATTACK

Definition of Extended Attack

Extended Attack is the suppression activity for a wildfire that has not been contained or controlled by initial attack or contingency forces and for which more firefighting resources are arriving, en route, or being ordered by the Initial Attack Incident Commander.

An Extended Attack Incident is the phase of the incident when Initial Attack capabilities have been exceeded. This has a high potential for more serious accidents and injuries. All planned actions must consider firefighter and public safety as the number one priority.

When complexity levels exceed Initial Attack capabilities, the appropriate ICS positions should be added to the command staff commensurate with the complexity of the incident. Complexity is usually Type 3; however complexity could be typed at any level.

Characteristics of an Extended Attack Incident

An Extended Response Incident is normally characterized by:

- Usually less than 100 acres in size; however, size is only one determining factor.
- Firefighting resources vary from several single resources to several Task Forces/Strike Teams.
- The incident may be divided into divisions, but would not meet the Division/Group Supervisor complexity.
- The incident is not expected to be contained and/or controlled in the first operational period.
- Generally, a written IAP would not be needed or prepared.
- Some of the Command and General Staff functions, such as Operations, Planning, Logistics, Safety, and Liaison, may be staffed.
- Staging Areas may be used and in some instances a small incident Base established.

Example of an Extended Attack Organization

General Staff positions filled as needed.

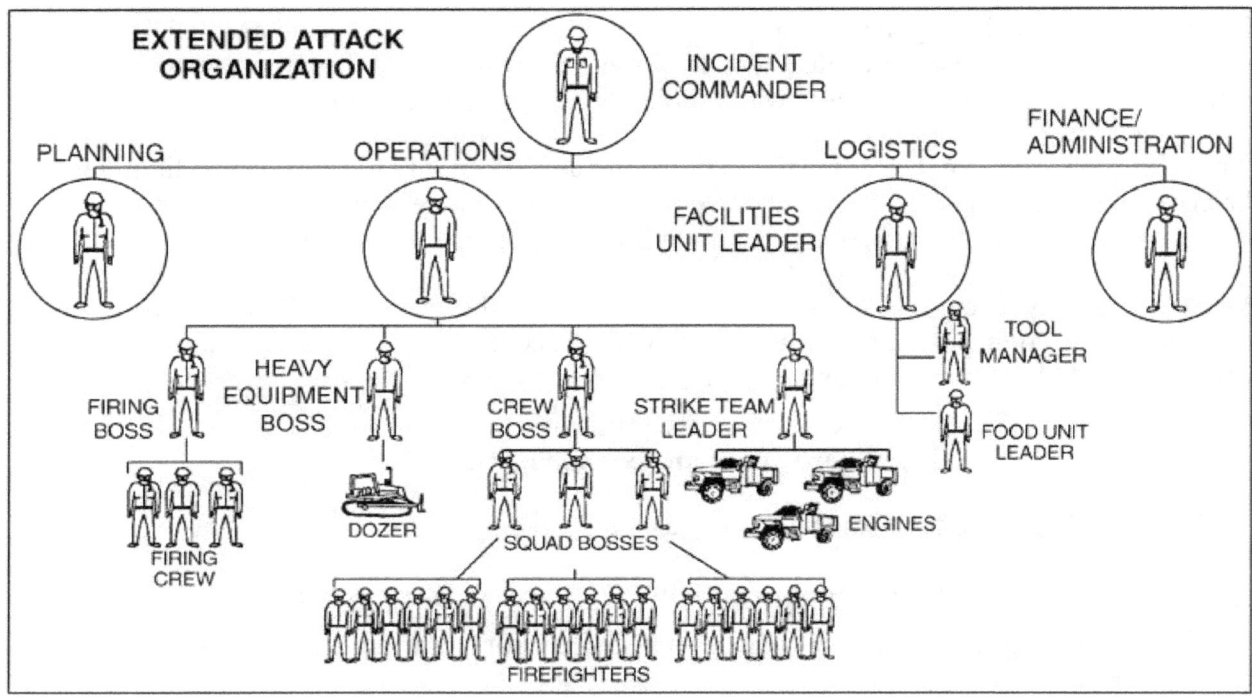

Change From an Initial Attack Incident to an Extended Attack Incident

Early recognition by the Initial Attack IC that the Initial Attack forces will not control a fire is important. As soon as the Initial Attack IC recognizes that additional resources are needed or knows additional forces are en route, the IC may need to withdraw from direct fireline suppression and must prepare for an Extended Attack mode of operation.

The Initial Attack IC will perform the following duties when changing to an Extended Attack Incident if all positions are not filled:

- Establish an Incident Command Post (ICP) and check-in location(s) to receive, brief, and assign incoming resources.

- Use the *Risk and Complexity Assessment* to validate organizational needs.

- Follow the risk management process in the *IRPG*. Review and update regularly during the incident.

- Employ strategy and tactics that will:
  - ✓ Follow the Standard Firefighting Orders.
  - ✓ Ensure Watch Out Situations are mitigated.
  - ✓ Ensure work/rest requirements are met.
  - ✓ Ensure entrapment situations are avoided.
- Determine and document incident objectives. Included in the objectives will be triggers or decision points for disengagement.
- Complete and document incident complexity.
  - ✓ Type 3 or greater complexity incidents require an IC without collateral duties, such as Logistics, Planning, or duty officer.
- Use an Incident Briefing Form (ICS 201) to:
  - ✓ Sketch a map of the fire and identify resource assignments.
  - ✓ Document the fire organization.
  - ✓ Keep track of all resources that are on scene, en route, and ordered.
  - ✓ Document strategy, tactics, and current actions.
- Review Extended Attack Safety Checklist.
- Keep Dispatch or other higher level officer informed of:
  - ✓ Status of the fire
  - ✓ Progress of the suppression effort
  - ✓ Additional resources needed
  - ✓ Weather conditions, especially changes
  - ✓ Special situations, such as values threatened, etc.
- As additional resources arrive:
  - ✓ Divide the fire into areas of responsibility, such as right and left flank or Division A and Division B.
  - ✓ Assign individuals responsibility for these areas. At first these will usually be Single Resources Bosses, but as multiple single resources arrive, consider aggregating them into Task Forces with a Task Force Leader to reduce span-of-control (recommended no more than 1:5) and increase suppression efficiency.
- As the incident continues to escalate, there may be a need to staff functional areas. These may be staffed by personnel at the unit leader level or by individuals who can complete the duties. Should the complexity require a fully qualified Section Chief, then the transition to a Type 2 Organization should begin. Assign the following as needed:
  - ✓ A person to directly supervise the suppression efforts.
  - ✓ A person to begin assessing logistical needs, such as feeding, fuel, sleeping arrangements, special equipment, etc.

Chapter 2 – Operational Guides

- ✓ A person to address incident planning needs:
    - Establish formal check-in and resource status.
    - Gather, record, and provide onsite information to firefighting personnel and Dispatch.
    - Take onsite weather check, and obtain weather reports and forecasts.
    - Start written IAP, if required by IC.
    - Prepare maps.
    - A Liaison Officer (LOFR) is especially important in multiple agency or jurisdiction incidents.
    - A Safety Officer (SOFR).

Control or Transfer to Type 2 Incident

At some point, the fire will be contained, controlled, or a decision will be made to transition to a larger, more complex organization.

Key indicators as to when to make this transition are:

- Incident objectives will not be met.
- The fire will not be controlled in the first or next operational period.
- A written IAP will be needed for the next operational period.
- Logistical support is needed, such as an Incident Base or Camps to feed, sleep, and supply personnel on the fire.
- There is a need to fill most or all of the Command and General Staff positions.
- Fire complexity exceeds capability of Extended Attack Organization.

If the Extended Attack IC follows the procedures identified above, the efficiency of the suppression action will be optimized and the fire will either be controlled or the stage will be set for a smooth transfer of command to the incoming Type 2 Organization.

The primary objective of all ICs is to provide for firefighter and public safety. Discharge of this objective applies the appropriate suppression response. This objective may require transfer of command. A measurable performance element with safety implications is the execution of this transfer of command. Adequate staffing, ordering of needed resources, good planning, good documentation, and quality briefings are all important elements of transfer of command.

Extended Attack Safety Checklist

After your initial sizeup of the fire and/or transition from an Initial Attack IC, answer the following questions (repeat this analysis whenever there is a change in conditions on the fire or a predicted change in fire conditions). If the answer is NO to any of the checklist questions, you MUST take corrective action(s) **IMMEDIATELY**.

Yes	No	Question
		Do you have a current fire weather forecast for the fire location?
		Is the observed fire weather consistent with the forecast?
		Can you control the fire with the resources available (on the incident or soon to be on the incident) under expected conditions?
		Have you developed a plan to attack the fire? Direct or indirect, anchor points, escape routes, head or flank attack or priority areas. Have you communicated the plan to all personnel assigned to the incident, including new arrivals?
		Are Lookouts in place or you can see all of the fire area?
		Can you communicate with everyone on the fire and with Dispatch?
		Escape routes are established. If you are using the black, is it completely burned and without a reburn potential?
		Safety and the Standard Firefighting Orders are being followed?
		Will you control the fire before the next operational period?
		Have you reported the status of the fire to Dispatch?
		Do you have a complete list of what resources have been ordered for the fire?
		Are cost-share issues present?
		Have all personnel on the fire been informed of the transition to an Extended Attack Incident and any change of plans?
		Fire complexity has exceeded management capability of Extended Attack Organization.
		Has this transition of command been documented in writing and through Dispatch?

Chapter 2 – Operational Guides

LARGE FIRE MANAGEMENT TEAMS

Type 2 Organization

A Type 2 Organization is the first level at which most or all of the Command and General Staff positions are activated and are filled by a Type 2 Incident Management Team (IMT). The IC and Command and General Staff must function as a team, handling many aspects, such as:

- Supervising a large organization.
- Planning during multiple operational periods.
- Gathering information to develop a written IAP.
- Providing logistical support, including the establishment and operation of a Base and possibly Camps.

Type 1 Organization

The primary difference between a Type 1 and Type 2 Organization is a matter of size and complexity. The factors that affect the decision to go to a Type 1 Operation are variable and depend to a large extent upon the needs and policies of the agency or agencies involved. The Type 1 Organization has all the characteristics of a Type 2 Organization plus:

- All Command and General Staff positions are filled with Type 1 qualified people.
- The number of divisions or groups may require that Branches be activated to address span-of-control needs.
- Operations personnel often exceed 500 people per operational period, and total personnel on the incident usually exceed 1,000.
- Aviation operations often involve several types and numbers of aircraft.

Organization Chart for Type 1 and Type 2 Incidents

Remember: Fill only those positions needed.

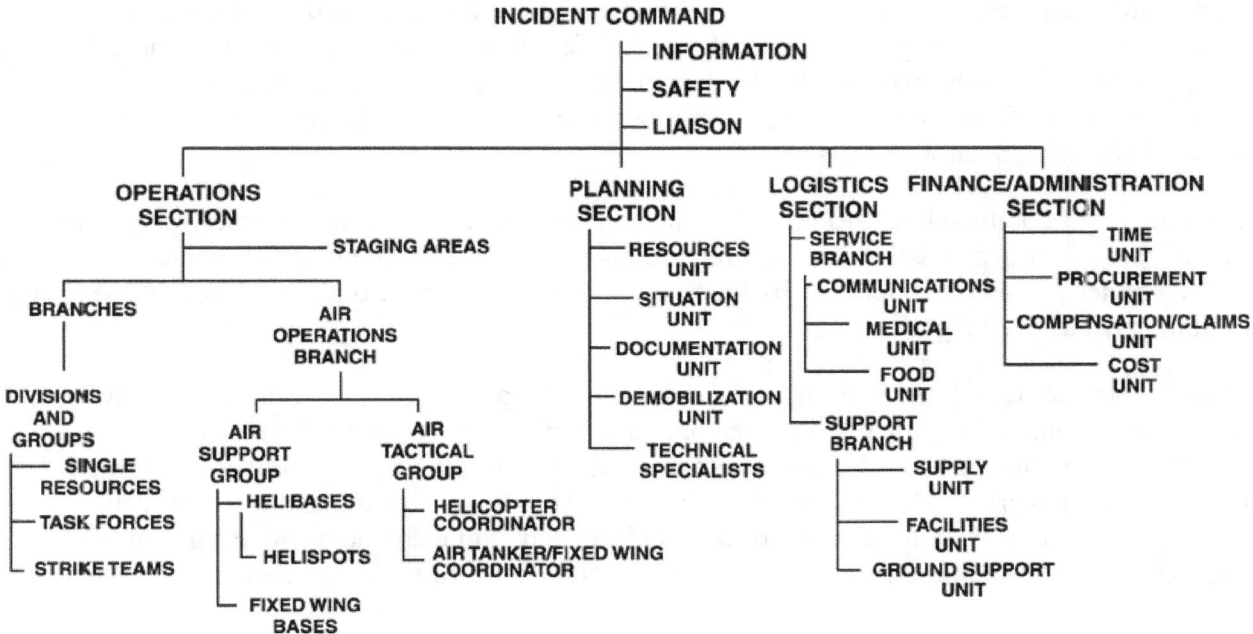

Area Command

Area Command is an expansion of the incident command function. It is designed to manage a very large incident that has multiple IMTs assigned. These teams may be established any time the incidents are close enough that oversight direction is required. This is to ensure that resource allocation conflicts do not arise among the IMTs.

The functions of Area Command are to coordinate:

- Objectives that conflict between incidents
- Strategies that conflict between incidents
- Priorities for the use of critical resources allocated to the incident or incidents assigned to the Area Command

The organization is normally small, with personnel assigned to Command, Planning, Aviation, and Logistics.

Unified Command

A representative from each of the involved agencies with jurisdiction authority shares command, and at times, other functions. Collectively, they direct the management of the incident to accomplish common objectives. Unified Command may be at the IMT or Area Command level.

TRANSFER OF COMMAND

A continuous command presence must be maintained on all incidents until all resources are released. Command of incidents, and some or all personnel in the incident management organization, may change one or more times as the incident changes in size or complexity, if it is of long duration or it changes jurisdiction(s). A briefing that captures all essential information for continuing effective command of the incident and provides for firefighter and public safety must occur before transfer of command. This information should be recorded and displayed for easy retrieval and subsequent briefings.

The transfer of command authorities for an incident must be as efficient and orderly as possible. The IC and his or her organization shall remain in charge of the incident until the incoming IC and his or her personnel are briefed by their counterparts and, where one is required, a delegation of authority has been signed.

Many safety problems emerge as an incident becomes larger and/or more complex. Incident transfer of command historically has been one of the most dangerous phases of incident management. Incidents should transfer command at a specific time, preferably at the start of a new operational period. The operational effort should continue during transfer period with command and control of the incident firmly in place, and with clear, achievable, and sound strategy and tactics communicated to and implemented by all firefighting resources.

Incident Commander Briefing

The outgoing Incident Commander must brief the incoming IC upon his or her arrival. The incoming IC should not assume command until thoroughly briefed and an exact time of command transfer is determined. If the incoming IC is arriving with a team, his or her team members may also attend the briefing. Likewise, if the outgoing IC has a team in place, those team members may also attend the briefing. After the briefing, incoming team members will start phasing into their areas of responsibility, but will not assume control until the predetermined time as agreed upon by the incoming and outgoing ICs. Notification of transfer of command must be immediately communicated to ALL firefighting resources, affected dispatch office(s), and Agency Administrator(s) through radio communication and/or verbal briefing.

Incident Commander's Checklist

The incoming IC, at all levels of complexity, should address the following items before he or she assumes command of an incident:

- Jurisdiction(s) responsible for the incident.
- Name, location, and radio contact of current IC(s).
- Agency Administrator(s) objectives for the incident.
- Current status of the incident and resources.
- Current map(s) of incident.
- Fuel and weather conditions; current, predicted, variations from normal (described in terms of expected fire behavior).
- Fire history of the incident area, including any unusual or potentially unusual fire behavior.
- Firefighter and public safety and safety concerns.
- Other agencies on the incident and their representatives.
- Transportation routes to and from the incident.
- Date and time for transfer of command.
- Primary contact for coordination and support.
- Radio frequencies assigned to incident.
- Necessary releases of any assigned resources.
- Reporting requirements (situation updates to Dispatch, Agency Administrator(s), ICS 209, etc.).
- Resource ordering protocols.
- Other (use of trainees, public information).

Agency Administrator(s)' Responsibility for the Transfer of Command and Release of Incident Management Teams

The following guidelines are for the orderly transfer of command of fire management authorities to incoming ICs and their teams as well as their release. Agency Administrator(s) always maintain responsibility for the incident. Some information will need to be in writing and some may be verbal.

Transfer of Authority

- The IC in place is in charge until officially released. Release should not occur until incoming IC and team members are briefed by their counterparts and ready to take full command of incident.
- The operational effort should continue during the transfer period, with command and control of the incident firmly in place, and with clear, achievable, and sound strategy and tactics communicated to and implemented by all firefighting resources. As a general rule, command transfer should occur at the end of an operational period.
- The requesting unit should specify the expected time of arrival and expected time of transfer of command to the incoming team.
- The current IC should contact the local Agency Administrator in advance for location and time for Agency Administration briefing.
- The requesting agency should accomplish the following before the arrival of the incoming team:
    - ✓ Make contact with incoming IC before his or her arrival. Give IC an update on progress of fire and inquire if there are any special needs for the team.
    - ✓ Determine ICP, Base, and Staging Area locations.
    - ✓ Order support equipment, supplies, and initial basic support organization for the incident.
    - ✓ Determine transportation needs of the team, and obtain needed vehicles.
    - ✓ Schedule the Agency Administrator briefing time and location.
    - ✓ Obtain necessary information for the Agency Administrator briefing.
    - ✓ Obtain necessary communications equipment and support for the incident.
- It is the responsibility of the jurisdictional Agency Administrator(s) to ensure that, where required, the Wildland Fire Decision Support System (WFDSS) is used.
- The existing IC at the ICP should brief the incoming IC and team. The time of transfer of command will depend upon incident complexity, expertise of the existing team, and/or other problems.
- Complete a written Delegation of Authority, per agency policy, for the incoming IC to review.

Agency Administrator Briefing

An Agency Administrator briefing should take place as soon as the incoming team is completely assembled.

Release of an Incident Management Team

The Agency Administrator must approve the date and time for the release of an IMT. The outgoing IC should start phasing in the incoming team members before demobilizing outgoing team members.

- The outgoing team should not be released from the incident until fire management activity and workload is at a level that the incoming team can reasonably assume. Some considerations to assist in this determination are:
    - ✓ A transfer of command plan should be prepared for the incoming IMT by the team being released.
    - ✓ Fire should be controlled or mopped up to a specified standard.
    - ✓ Unneeded resources have been released.
    - ✓ Base or Camp is reduced or being shut down.
    - ✓ Planning Section Chief has prepared a final copy of the fire report and narrative.
    - ✓ Finance/Administration Section Chief should have known finance problems resolved. Contact should be made with agency fiscal personnel.
    - ✓ Resource rehabilitation work is completed or to a point where the agency is satisfied with assuming remaining work.
    - ✓ Overhead performance ratings are completed.
- The departing team should have an internal debriefing session before meeting with the Agency Administrator.
- The Agency Administrator should debrief the departing team and prepare a written evaluation as soon as possible after release, according to agency policy.

URBAN INTERFACE

Wildland/Urban Interface "Watch Out" Situations

- Poor access and narrow, one-way roads
- Bridge load limits
- Wooden construction and wood shake roofs
- Power lines, propane tanks, and HazMat threats
- Inadequate water supply
- Natural fuels 30 feet or closer to structures
- Structures in chimneys, box canyons, narrow canyons, or on steep slopes (grade 30% or more)
- Extreme fire behavior
- Strong winds
- Evacuation of public (panic)
- Underground utilities threat
- Structural collapse zone when structures are exposed to fire
- Smoke byproducts often laced with chemical compounds not found in pure wildland fires

Structures exposed to wildland fire in the urban interface can and should be considered as another fuel type. Sizeup and tactics should be based upon fuels, weather, and topography, just as they would be applied to a wildland fire.

Identification of Reduced-Risk Structures and Communities

Ask local fire departments, emergency management offices, and/or law enforcement agencies to what extent their communities are fire adapted. Fire adapted communities have undertaken strategies to mitigate fire risk so that should a wildfire impact a community, the community sustains minimal damage. Fire-adaption strategies include Firewise Communities and "Ready, Set, Go!" programs, as well as other individual and community actions to reduce risk. Request copies of plans and maps to assist in planning and preparing for protecting the community from wildland fire.

Structure Triage Guidelines

Defensible – Prep and Hold

- Determining Factor: Safety zone present.
- Sizeup: Structure has some tactical challenges.
- Tactics: Firefighters are needed onsite to implement structure protection tactics during fire front contact.

Defensible – Standalone

- Determining Factor: Safety zone present.
- Sizeup: Structure has very few tactical challenges.
- Tactics: Firefighters may not need to be directly assigned to protect the structure as it is not likely to ignite during initial fire front contact. However, no structure in the path of a wildfire is completely without need of protection. Patrol after the passage of the fire front will be needed to protect the structure.

Non-Defensible – Prep and Leave

- Determining Factor: NO safety zone present.
- Size up: Structure has some tactical challenges.
- Tactics: Firefighters are not able to commit to stay and protect the structure. If time allows, rapid mitigation measures may be performed. Set a trigger point for a safe retreat. Remember preincident preparation is the responsibility of the homeowner. Patrol after the passage of the fire front will be needed to protect the structure.

Non-Defensible – Rescue Driveby

- Determining Factor: NO safety zone present.
- Size up: Structure has significant tactical challenges.
- Tactics: Firefighters are not able to commit to stay and protect the structure. If time allows, check to ensure that people are not present in the threatened structure (especially children, the elderly, and invalids). Set a trigger point for a safe retreat. Patrol after the passage of the fire front will be needed to protect the structure.

Structure Assessment Checklist (if Time Permits)

Address and Property Name

- Numerical street address, ranch name, etc.
- Number of residents onsite

Road Access

- Road surface (paved, gravel, unimproved, dirt)
- Adequate width; vegetation clearance and safety zones along road
- Undercarriage problems (4x4 access only)
- Turnouts and turnarounds
- Bridges (load limits)
- Stream crossings (approach angle, crossing depth, and surface)
- Terrain (road slope and location on slope – near chimneys, saddles, canyon bottom)
- Grade (more than 15%)

Structure or Building

- Single residence, multicomplex, or outbuilding (barn or storage)
- Does the building have unknown or hazardous materials?
- Exterior walls (stucco or other noncombustible, wood frame, vinyl, wood shake)
- Large, unprotected windows facing heat source
- Proximity of any aboveground fuel tanks (liquefied petroleum gas (LPG), fuel oil, etc.)
- Roof material (wood shake, asphalt, noncombustible)
- Eaves (covered with little overhang, or exposed, with large overhang)
- Other features (wood deck, wood patio cover and furniture, wood fencing)

Clearances, Exposures, and Defensible Space

- Structure location (narrow ridge, canyon, midslope, chimney)
- Adequate clearance around structure (minimum of 100 feet; the steeper the slope, the more clearance is required)
- Surrounding fuels (the larger and denser the fuels, the more clearance required)
- Flammable fuels (trees, ladder fuel, shrubs) adjacent to structure (is there time for removing these fuels?)
- Other combustibles near structure (wood piles, furniture, fuel tanks)

- Is there adequate clearance around fuel tank?
- Power lines or transformers (DO NOT park under power lines)

Hazardous Materials
- Chemicals (look for DOT, NFPA, or UN symbols)
- Pesticides and herbicides
- Petroleum products
- Paint products

Water Sources
- Hydrant or standpipe (When connecting with hydrant, be aware of flow rate and gal/min output; size and venting capability of engine or water tender may not be able to handle hydrants with high flow and gal/min rates.)
- Storage tank
- Swimming pool
- Hot tub
- Fish pond
- Irrigation ditch

Evacuation
- Is safe evacuation possible? (Identify safe refuge for those who cannot be evacuated.)
- Coordinate with onscene law enforcement and emergency services personnel.

Estimated Resources for Protection
- Number(s) and type(s) of engines, water tenders, crews, dozers (General guidelines: One engine per structure, one additional engine for every four structures to be used as "backup" and for patrol. For structures that are close together (50 feet or less), one engine may be adequate to protect two structures.)
- Type and number of aircraft available

Structure Protection Guidelines

DO NOT COMMIT TO STAY and protect a structure UNLESS A SAFETY ZONE FOR FIREFIGHTERS AND EQUIPMENT HAS BEEN IDENTIFIED AT THE STRUCTURE during sizeup and triage. Move to the nearest safety zone, let the fire front pass, and return as soon as conditions allow.

Placing Equipment

- If visibility is zero due to smoke from the fire, do not try to move your vehicles, as many accidents have occurred this way. Stay in your safety zone behind the structure, or use the structure as a refuge.
- Identify escape routes and safety zones, and make them known to all crewmembers.
- ALWAYS STAY MOBILE, and wear all of your PPE.
- Back the equipment in for a quick escape.
- Mark the entrance to long driveways to show that protection is in place (**very important** when the structure cannot be seen from the road). Always notify Operations Section of the meaning of markings used:
    - ✓ Ribbon (flagging) across the drive entrance
    - ✓ Multiple ribbons on the street at the end of a drive
    - ✓ A sign
    - ✓ Other predetermined markings
- Park in a cleared area (watch for overhead hazards).
- Protect your equipment (park behind the structure, placing the structure between the equipment and the fire front; be aware of spot fires occurring behind you).
- Watch for hazards (dropoffs, potholes, aboveground fuel storage, chemicals, septic tanks).
- Keep egress route clear:
    - ✓ Park extra equipment on the street.
    - ✓ Keep all hoses off the driveway.
- Have an engine and/or crew protection line charged and readily available.
- DO NOT make long hose lays.
- Try to keep sight contact with all crew members.

Water-Use Guidelines
- Keep at least 100 gallons of water reserve in your tank.
- Top off tank at every opportunity; use a garden hose.
- Draft water from a swimming pool, hot tub, and/or fishpond.
- STAY MOBILE. Do not hook up to the hydrant except to refill the tank. (Hydrant may not always work if the system is electric powered and power is lost in the area.)
- CONSERVE WATER; avoid wetting down an area.
- Apply water only if it controls fire spread or significantly reduces heating of the structure being protected.
- Keep fire out of the heavier fuels.
- Extinguish fire at its lowest intensity, not when it is flaring up.
- Knock down fire in the lighter fuels.
- Have enough water to last the duration of the main heat wave and to protect the crew.
- Apply compressed air foam (CAF) or gel, if available.

Preparing the Structure
- Determine if residents are home (legal responsibility for evacuation lies with law enforcement). If residents remain on scene, advise them to use the structure, if it's safe to do so, as a refuge when the fire arrives.
- Clean leaves, needles, and any other combustible materials (in accordance with agency policy) off of the roof.
- Cover vents and any air conditioning unit(s) on the roof (in accordance with agency policy).
- Remove and scatter away from the structure:
  - ✓ Overhanging limbs
  - ✓ Ground or ladder fuels to prevent fire from moving into the crowns
  - ✓ Wooden fences and wood piles near the structure
- Clear the area around any aboveground fuel tank(s), and shut off the tank(s).
- Place combustible outside furniture inside the structure.
- Close windows and doors, including the garage door, leaving all of them unlocked. **As a last resort, you may need to use the structure as a refuge.**
- Have garden hoses charged, and place them strategically around the structure for immediate use.

Building Refuge

Seeking refuge in a building or structure is an option supervisors may want to consider for crew protection when a change in fire behavior prevents reaching an escape route or safety zone. Agency guidelines MUST be considered when deciding to use a building or structure as crew protection.

- Advise your immediate supervisor (Strike Team Leader, Division/Group Supervisor, etc.) of the situation.
- If time allows, remove combustible materials (lawn furniture, wood piles, etc.) and vegetation away from the structure and any propane tank(s).
- Close windows and heavy drapes. Take down light-weight curtains, and close exterior doors.
- Bring into the structure fire extinguishers, back pumps, and charged hose line, if available.
- Fill all sinks, bathtubs, and any available buckets with water, and soak towels and other heavy cloth items to place against exterior door jams.
- KEEP AWAY from windows and exterior doors as the fire passes.
- STAY OUT of the basement and upper floors.

CHAPTER 3 – POSITION RESPONSIBILITIES

All positions within an ICS Organization share some common responsibilities. These are enumerated in the Federal Emergency Management Agency (FEMA) *Emergency Responder Field Operating Guide* (*ERFOG*). It is assumed that all individuals will be familiar with the common responsibilities. Among those are coming to the incident prepared with necessary work materials, receiving a briefing from the immediate supervisor, and documenting your activities in an Activity Log (ICS 214). For additional information on common responsibilities, refer to the *ERFOG*.

Specific responsibilities for each position are presented below, along with organization charts and position checklists.

COMMAND AND GENERAL STAFF

Organization Chart

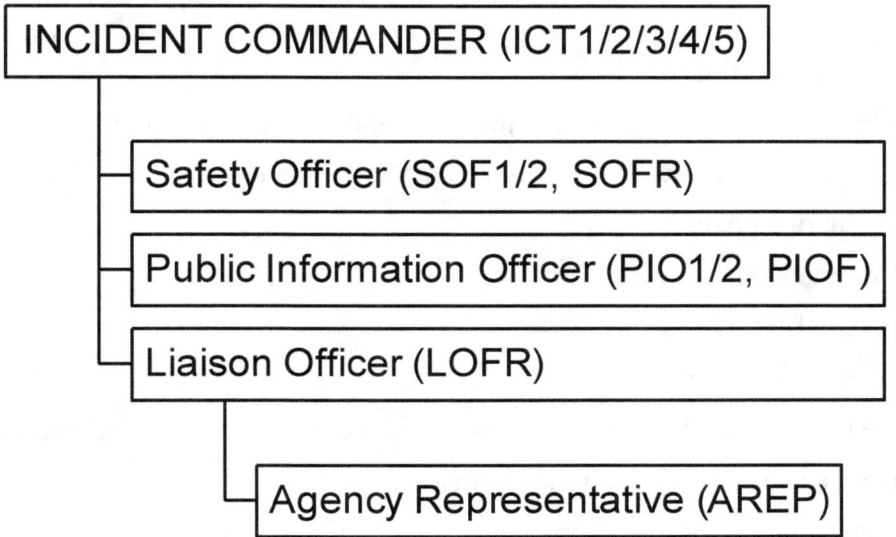

Chapter 3 – Position Responsibilities

Position Checklists

Incident Commander (ICT1/2/3/4/5)

The Incident Commander is responsible for all incident activities.

Critical Safety Responsibilities

- Ensure that safety receives priority consideration in the analysis of strategic alternatives, the development of the IAP, and in all incident activities.
- Assess incident situation, both immediate and potential.
- Conduct risk assessment for all strategic alternatives.
- Maintain command and control of the incident management organization.
- Ensure that the safety and welfare of all incident personnel and the public are maintained.
- Ensure transfer of command is announced to host unit Dispatch and to all incident personnel.

Other Duties

- Review Common Responsibilities
- Obtain briefings from Agency Administrator and/or prior Incident Commander.
- Obtain Delegation of Authority from Agency Administrator.
- Set incident objectives.
- Brief Command and General Staff.
- Approve the IAP.
- Determine information needs.
- Approve requests for additional resources and requests for release of resources.
- Approve the use of trainees on the incident.
- Authorize release of information to news media, if delegated by Agency Administrator.
- Ensure Incident Status Summary (ICS 209) is completed and forwarded to agency Dispatch Center(s) on schedule.
- Approve Demobilization Plan.
- Conduct strategy meetings, reviewing, validating, and/or revising the WFDSS, incident objectives, strategies, and tactics.
- Determine effects of control actions on environmental and ecological processes.
- Ensure that strategic and tactical options consider all resource values.
- Foster an atmosphere free of discrimination, sexual harassment, and other forms of inappropriate behavior.

- Supervise staff activities, ensure functional performance is maintained, and take corrective action, if needed.
- Participate in external incident affairs as required.
- Ensure incident financial accountability and expenditures are maintained to agency standards.
- Ensure incident documentation package is complete.
- Debrief with Agency Administrator.

Safety Officer (SOF1/2, SOFR)

The Safety Officer, a member of the Command Staff, is responsible for monitoring and assessing hazardous and unsafe situations and developing measures for assuring personnel safety. The Safety Officer will correct unsafe acts or conditions through the regular line of authority, although they (Safety Officer) may exercise emergency authority to stop or prevent unsafe acts when immediate action is required.

Only one Safety Officer will be assigned for each incident. The Safety Officer may have assistant Safety Officers as necessary, and the assistant Safety Officer may represent assisting agencies or jurisdictions. Assistant Safety Officers may have specific responsibilities, such as air operations, hazardous materials, etc.

Critical Safety Responsibilities

- Analyze proposed and selected strategic alternatives from a safety perspective, ensuring that risk management is a priority consideration in the selection process.
- Direct intervention will be used to immediately correct a dangerous situation.
- Prepare the safety message included in the IAP.
- Develop the Incident Action Plan Safety Analysis (ICS 215A) planning matrix with the Operations Section Chief.
- Present safety briefing to overhead. Safety briefing should emphasize hazards and risks involved in action plan components.

Other Duties

- Establish systems to monitor fire activities for hazards and risks. Take appropriate preventive action.
- Priority of recommendations will start with risks having the highest potential for death or serious injury and follow through to those of lesser degree.
- Establish operating procedures for assistant Safety Officers.
- Evaluate operating procedures. Update or modify procedures to meet the safety needs on the fire.
- Review and approve Medical Plan (ICS 206).

- Review IAPs to ensure that safety issues have been identified and mitigated.
- Analyze observations from staff and other personnel.
- Ensure accidents are investigated.
- Prepare accident report upon request of the Incident Commander.
- Monitor operational period lengths of incident personnel to ensure work/rest guidelines are followed; recommend corrective action to Incident Commander.
- Monitor food, potable water, and sanitation service inspections. Request assistance from health departments as needed.
- Monitor incident PPE needs.
- Inspect incident facilities, hand tools, power equipment, vehicles, and mechanical equipment.
- Monitor driver or operator qualifications and operational periods.
- Monitor all air operations; review aircraft incidents and accident reports.
- Ensure appropriate accident, incident, and other safety reports (such as SAFECOMs and SAFENETs) are completed and submitted.
- Prepare final safety report upon request of the Incident Commander.

Public Information Officer (PIO1/2, PIOF)

The Public Information Officer, a member of the Command Staff, is responsible for the formulation and release of information about the incident to the news media, local communities, incident personnel, other appropriate agencies and organizations, and for the management of all Public Information Officers assigned to the incident.

- Contact the jurisdictional agency to coordinate public information activities.
- Obtain copies of current Incident Status Summaries (ICS 209).
- Develop policy with Incident Commander, Agency Administrator, agency Public Affairs Officer, IMT members, and incident investigators regarding information gathering and sharing. Observe constraints on release of information.
- Develop and receive Incident Commander's approval of a comprehensive, proactive communications strategy that reflects both immediate and long-term goals.
- Prepare initial information summary as soon as possible after arrival.
- Obtain approval for release of information from Incident Commander.
- Attend meetings to update information releases.
- Arrange for meetings between media and incident personnel.
- Provide escort service to the media and very important persons (VIPs); provide PPE as necessary.
- Respond to special requests for information.

- Keep informed of incident developments and control progress through Planning Meetings and regular contacts with other incident staff, host unit, and cooperating agencies.
- Keep the Incident Commander informed of any potential issues involving the general public, news media, or other sources.

Liaison Officer (LOFR)

The Liaison Officer, a member of the Command Staff, is the point of contact for the assisting and cooperating Agency Representatives. This includes Agency Representatives from other fire agencies, Red Cross, law enforcement, public works, etc.

- Provide a point of contact for assisting and cooperating Agency Representatives.
- Identify each Agency Representative, including communications link and location.
- Maintain a current list of cooperating and assisting agencies assigned. Confirm resource list with the Resource Unit Leader.
- Respond to requests from incident personnel for interorganizational contacts.
- Monitor incident operations to identify current or potential interorganizational problems.
- Remain visible on the incident to incoming cooperators and assisting agencies.
- Respond to requests for information, and resolve problems.
- Participate in Planning Meetings providing current resource status, limitations, and capability of other agency resources.
- Keep cooperating and assisting agencies informed of planning actions.

OPERATIONS

Organization Chart

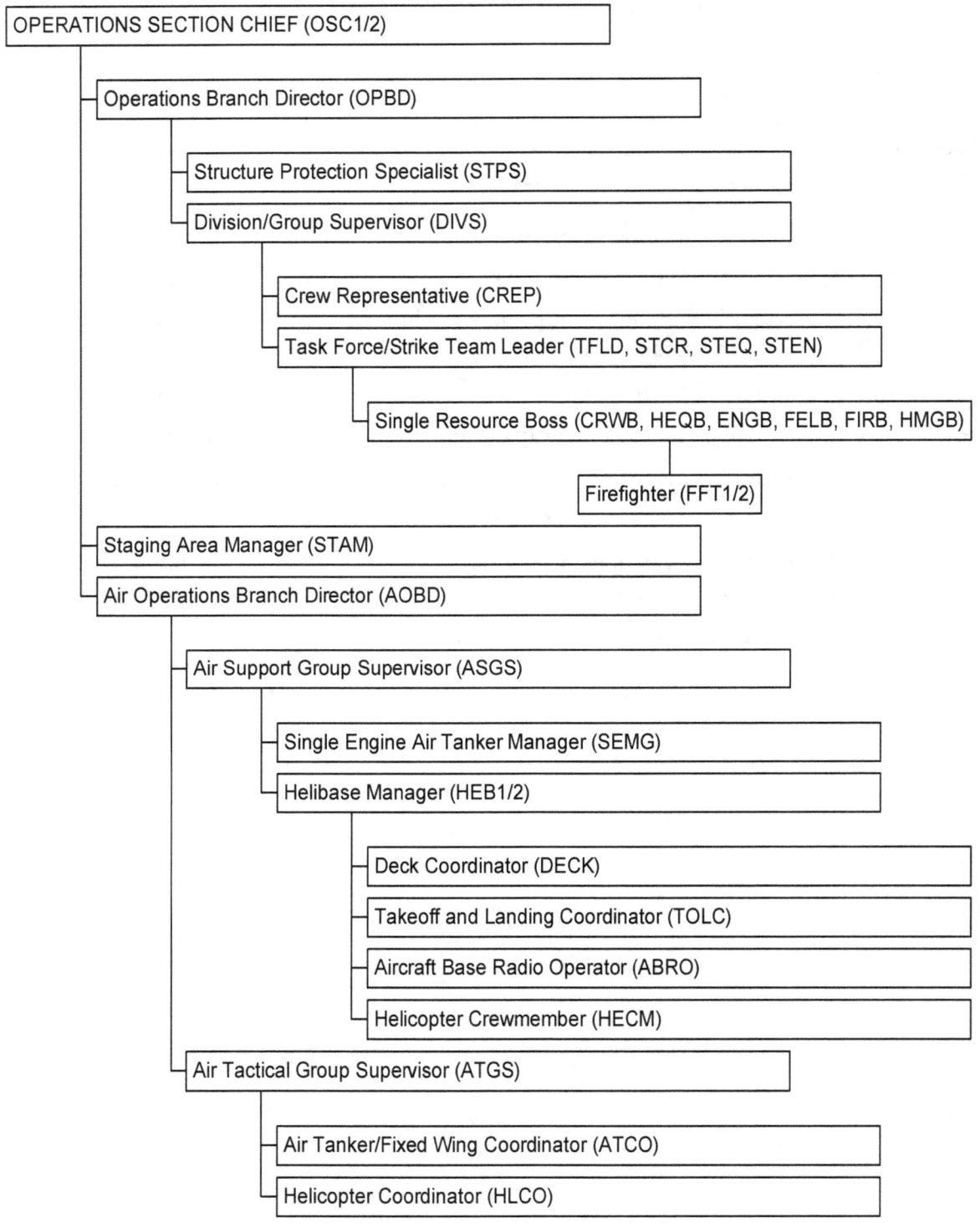

Position Checklists

Operations Section Chief (OSC1/2)

The Operations Section Chief, a member of the General Staff, is responsible for managing all operations directly applicable to the primary mission.

Critical Safety Responsibilities

- Use the risk management process, and supervise operations.
- Maintain accountability of assigned resources.
- Develop Incident Action Plan Safety Analysis (ICS 215A) with Safety Officer.

Other Duties

- Develop operations portion (ICS 215) of the IAP with the Planning Section Chief.
- Brief and assign Operations personnel according to the IAP.
- Facilitate and coordinate the ordering and release of operation resources.
- Assemble and disassemble Task Forces/Strike Teams and assigned to operations.
- Report special activities, events, and occurrences to Incident Commander.

Operations Branch Director (OPBD)

The Operations Branch Director, when present, is responsible for implementing the portion of the IAP applicable to the assigned Branch.

Critical Safety Responsibilities

- Use the risk management process and supervise Branch operations.
- Maintain accountability of assigned resources.
- Provide safety briefing to subordinate resources.

Other Duties

- Attend Planning Meetings at the request of the Operations Section Chief.
- Review Division/Group Assignment Lists within the Branch.
- Brief and assign specific work tasks to Division/Group Supervisors.
- Approve accident and medical reports.
- Resolve logistic problems reported by subordinates.

Structure Protection Specialist (STPS)

The Structure Protection Specialist, when activated, is primarily responsible for preparing and protecting structures threatened by wildfire. The STPS may work directly for the Operations Section Chief or may be assigned to an Operations Branch Director or Division/Group Supervisor.

Critical Safety Responsibilities

- Use the risk management process while planning structure protection activities.
- Maintain accountability of assigned resources.
- Provide safety briefing to subordinate resources.
- Coordinate activities with adjacent Branches and/or Divisions.
- Keep supervisor informed of situation and resources status.

Other Duties

- Help coordinate activities with local municipal firefighters.

Division/Group Supervisor (DIVS)

The Division/Group Supervisor is responsible for implementing the assigned portion of the IAP.

Critical Safety Responsibilities

- Use the risk management process, and supervise operations in the Division.
- Maintain accountability of assigned resources at all times.
- Coordinate activities with adjacent Divisions.
- Keep supervisor informed of situation and resources status.
- Provide safety briefing to subordinate resources.

Other Duties

- Brief and assign specific work tasks to Task Forces/Strike Team Leaders.
- Inform Incident Communications of all status changes of assigned resources.
- Ensure that assigned personnel and equipment get on and off the fireline in a timely and orderly manner.
- Resolve logistics problems within the Division/Group.
- Approve and turn in time for all resources in Division/Group to the Time Unit.

Crew Representative (CREP)

A Crew Representative may be provided by the sending agency for each hand crew sent to a fire. The Crew Representative is responsible for the welfare of the crew and provides a contact between the crew and the appropriate Incident Command Organization.

Critical Safety Responsibilities

- Maintain communications between the crew and the appropriate supervisors regarding the crew's safety and welfare.
- Provide safety briefing to assigned crew.

Other Duties

- Look after the crew's welfare on and off the fireline.
- Report crew status to assigned supervisor.
- As needed, maintain contact with crew's home Base.
- Report the crew's performance and problems to the sending agency's headquarters upon completion of the assignment.
- Coordinate with Interagency Resource Representative (IARR) if assigned.
- Participate in After Action Reviews (AARs) when appropriate.

Task Force/Strike Team Leader (TFLD, STCR, STEQ, STEN)

The Task Force/Strike Team Leader reports to a Division/Group Supervisor and is responsible for performing tactical missions as assigned on a division or segment of a division. The Leader reports work progress, resource status, and other important information to his or her supervisor and maintains work records on assigned personnel.

Critical Safety Responsibilities

- Use the risk management process, and supervise operations.
- Maintain accountability of assigned resources.
- Provide safety briefing to subordinate resources.
- Coordinate activities with adjacent resources.

Other Duties

- Brief and assign specific work tasks to assigned resources.
- Travel to and from the fireline with assigned resources.
- Retain control of assigned resources while off the fireline (feeding, timekeeping, sleeping area assignment, etc.).
- Turn in time for resources to Division/Group Supervisor.
- Evaluate performance of subordinates.

Single Resource Boss (CRWB, HEQB, ENGB, FELB, FIRB, HMGB)

A Single Resource Boss is responsible for supervising and directing a fire suppression module, such as a hand crew, engine, helicopter, heavy equipment, firing team, or one or more fallers.

Critical Safety Responsibilities

- Use the risk management process, and supervise operations of the resource.
- Maintain accountability of assigned resource.
- Provide safety briefing to subordinate resources.
- Coordinate activities with adjacent resources.

Other Duties

- Brief and assign specific work tasks to assigned resource.
- Retain control of assigned resource while off the fireline (feeding, timekeeping, sleeping area assignment, etc.).
- Turn in time for the resource(s) used to supervisor or Task Force/Strike Team Leader.
- Evaluate performance of subordinates.
- Return equipment and supplies to appropriate unit.

Firefighter (FFT1)

A FFT1 (Squad Boss) is a working leader of a small group (usually not more than seven members) and is responsible for keeping assigned personnel fully employed on assigned jobs. Normally this position is supervised by a Single Resource Crew Boss but may be assigned independently on occasion.

Critical Safety Responsibilities

- Ensure that instructions are clear and understood. Ask if you don't know!
- Know your skill level and limitations. Use the risk management process when accepting assignments.
- Keep supervisor informed on progress of assignment.
- Report any changes in fire behavior.
- Report all accidents, injuries, or hazardous conditions to supervisor.
- Wear your PPE.
- Ensure personnel have proper tools and know how to care for and use them.
- Look after the safety of assigned personnel.

Other Duties

- Ensure that personnel have water and lunches.
- Keep time when requested by supervisor.
- Report problems with personnel to supervisor.

Firefighter (FFT2)

A firefighter is the basic resource used in the control and extinguishment of wildland fires and works either as an individual or as a member of a crew under the supervision of a higher qualified individual.

Critical Safety Responsibilities

- Ensure that instructions are clear and understood. Ask if you don't know!
- Know your skill level and limitations. Use the risk management process when accepting assignments.
- Report any changes in fire behavior or hazardous conditions to supervisor.
- Report all accidents or injuries to supervisor.
- Wear your PPE.

Other Duties

- Perform manual and semiskilled labor as assigned.
- Participate in AARs when appropriate.
- Look, listen, and learn. Ask questions when appropriate.

Staging Area Manager (STAM)

A Staging Area Manager is responsible for managing all activities within a Staging Area.

Critical Safety Responsibilities

- Maintain Staging Area in safe operating condition.
- Post traffic plan for the Staging Area.

Other Duties

- Establish layout for Staging Area.
- Determine and order support needed.
- Establish check-in function as needed.
- Respond to requests for resource assignments.
- Report resource status changes as required.

Air Operations

Incident Operations – Aircraft

Aircraft are used for tactical and logistical needs. Aircraft can be effective tools for ICs, but aircraft only support ground-based operations. Tactical operations plans should not rely solely on aircraft for success; environmental conditions, fuel, and mechanical systems can impede aircraft operations. Aircraft can provide the IC the capability to:

- Deliver firefighters (crews, helitack, smokejumpers) to remote areas
- Deliver equipment
- Deliver water and/or fire retardant in support of ground operations
- Provide incident airspace management
- Provide the IC or Operations with fire information about fire area; behavior; threats to property, the public, or firefighters; access; and progress of operations

Incident Command System (ICS)

Aircraft operations are within the Air Operations Branch. If there is not an established Operations Branch, then Air Operations (aircraft and supporting personnel) are under the IC. The ICS designates aircraft by ICS typing that derives from the aircraft capability. The standard types are Type 1, 2, 3, and 4 for helicopters and air tankers. Aerial supervision, reconnaissance, and utility/logistics aircraft are not "typed."

As the incident complexity increases, the type and scale of the Air Operations adjusts with various functions and personnel to manage the functions (see Position Checklists and Operations Organization Chart). Many aviation support functions (i.e., Air Tanker Base) are regional and are not directly supervised by a specific IC or Air Operations Branch Director (AOBD).

Risk Management

All aircraft operations must take into account the risk to pilot, aircrews, and ground personnel before committing to a flight. Factors that are considered are:

- The operational environment
- Appropriate aircraft being used
- The qualifications of the pilot
- Recognition of flight hazards (wires, towers, other aircraft, terrain, visibility, weather). Generally, effective aerial fire operations are limited by winds exceeding 30 to 35 mph.
- Mitigation of the hazards, communication, and coordination between pilot and ground personnel.

Density Altitude (DA) Limitations

All aircraft performance is negatively impacted by increasing temperature and altitude. Payloads are decreased and takeoff and landing requirements are increased. High DA situations also could have a negative impact on fuel duration and increase the amount of refueling stops.

Policy

All aircraft are operated within Federal law. The Federal Aviation Administration (FAA) regulates aviation operations. Some Federal, state, county, and city agencies own and operate aircraft. These are known as "public aircraft." Generally, they are operated within the FAA regulations. Most federally operated fire aircraft are contracted and adhere to Department of Interior (DOI) or U.S. Forest Service (USFS) policy (both policies are similar). Many states also contract aircraft and have state aviation operations policies. Most aerial firefighting is conducted during daylight hours. There are some exceptions for some state, county, and city "public aircraft."

Flight and duty policy is established to mitigate pilot fatigue. A pilot is limited to a maximum of 8 hours of flight time per duty period.

For example, USFS and DOI policy also limit flight hours to a cumulative 36 hours in any 6 days. There is a minimum hours of "off duty" time required before resuming "on duty." USFS and DOI policy also requires 2 days off in any 14-day period and 1 day off if the cumulative 36 flight hours are exceeded (maximum is 42 hours in any 6 days).

Incident Airspace Coordination

All wildfire incidents have a "Fire Traffic Area" (FTA) established. A typical FTA extends to a 5 mile horizontal radius and up to 2,500 feet above ground level (AGL). The purpose is to provide a safe flying environment and establish standard procedures. If there is no aerial supervision (Air Tactical Group Supervisor [ATGS], Aerial Supervision Module [ASM], or Lead plane) over the fire, incident aircraft will coordinate with the IC and with other aircraft. Typical operating altitudes for the FTA are:

- Helicopters – surface to 500 feet.
- Air tankers/Single Engine Air Tankers (SEATs) – 1,500 feet orbit and 1,000 feet for setting up for the retardant drop.
- Aerial supervision "Air Attack" – 2,500 feet and above.
- Lead/ASM operates similar to the air tankers.

Military Airspace

Some military airspace can significantly impact aerial firefighting operations if coordination with the controlling military authority is not done. Restricted airspace requires permission to enter.

Temporary Flight Restriction (TFR)

Incident Commanders can request from the FAA a TFR in order to provide a safe flying environment for incident aircraft. TFRs help provide safer incident airspace, but are not guarantees that nonincident aircraft will enter a TFR. Law enforcement aircraft can enter without permission, and media aircraft can enter after requesting entrance from the controlling authority of the TFR. A TFR cannot close an airport and prevent airport traffic. A standard TFR's dimensions are 5 nautical mile radius and 2,000 feet above the typical incident terrain.

Incident Aircraft Communications

Positive coordinated communications with aircraft are critical to safety and effective fire suppression operations. Standardized terminology and a good Operations Plan lead to effective target description, which minimizes the amount of confusion and time that aircraft spend in the "low and slow" flight profile.

There are four main fire aviation radio communications functions. They are:

- Air to air, which is used for coordination between aircraft
- Air to ground, which is used for coordination between the ground personnel and the aircraft
- Flight following, which is used to track and coordinate between the Dispatch Center and the aircraft
- Emergency guard, which is used for emergency and call up if all other channels fail

Position Checklists

Air Operations Branch Director (AOBD)

The Air Operations Branch Director reports to the Operations Section Chief and is primarily responsible for preparing the air operations portion of the IAP, for implementing its strategic aspects, and for providing logistical support to aircraft operating on the incident.

Critical Safety Responsibilities

- Obtain briefing from Operations Section Chief.
- Request declaration (or cancellation) of TFR.
- Coordinate airspace with other incidents and local or regional airspace coordinators.
- Apply risk management practices to all aviation operations.
- Ensure that agency aviation policies are established and followed.
- Establish procedures for emergency reassignment of aircraft on the incident.
- Inform the aerial supervisor of the air traffic situation external to the incident.

Other Duties

- Participate in preparation of the IAP.

- Provide IAP and Air Operations Summary Worksheet (ICS 220) to the Air Support Group and Fixed-Wing Air Bases.

- Determine coordination procedures and coordinate with appropriate Operation Section and Logistics Section personnel (Branch, Division, etc.). Coordinate incident aircraft support functions with the Air Support Group Supervisor (ASGS). Coordinate the incident aircraft tactical operations with the Air Tactical Group Supervisor (ATGS).

- Orders and releases incident aircraft as needed.

- Supervise all Air Operations activities associated with the incident.

- Schedule approved flights of nonincident aircraft in the restricted airspace area.

- Develop Incident Aircraft Mishap Response Plan, and coordinate mishap reporting with Agency Administrators' aviation management personnel and/or local dispatch unit.

Air Support Group Supervisor (ASGS)

The Air Support Group Supervisor reports to the Air Operations Branch Director (AOBD) and is responsible for planning and oversight of incident aircraft support functions (helibase, helispot and Fixed Wing Air Bases).

Critical Safety Responsibilities

- Obtain assigned helibase air-to-ground and deck operations frequencies from Communications Unit Leader or Incident Radio Communications Plan (ICS 205).

- Obtain appropriate crash-rescue service for helibases and helispots.

Other Duties

- Participate in Air Operations planning activities.

- Brief the Helibase Manager and Fixed-Wing Base Managers on daily incident plans.

- Request special air support items from appropriate sources through Logistics Section.

- Identify helibase locations and assist Operations Section with identifying suitable helispot locations.

- Work with Finance in land use agreements with landowners for incident aircraft Bases.

- Inform Air Operations Branch Director of special aircraft and/or pilot restrictions.

Single Engine Air Tanker Manager (SEMG)

The Single Engine Air Tanker Manager reports to the Fixed-Wing Base Manager or Air Support Group Supervisor, if assigned to an Incident Management Team.

Critical Safety Responsibilities

- Conduct pre-use and daily briefing with pilot and support crew.
- Regulate all aircraft and ground traffic on and around SEAT base of operation.
- Verify correct communications and frequency procedures are followed.
- Suspend operations due to safety issues or other appropriate concerns.

Other Duties

- Conduct pre-use walk-around inspection of aircraft and ground support equipment.
- Order aircraft services as provided in contract specifications.
- Perform as liaison with airport or airstrip management.
- Perform as liaison between the SEAT vendor and the user agency.
- Initiate and sign correspondence and other contract administration documents.
- Complete all required forms, records, reports, and documents as required by using agency.
- Record and approve availability and flight times.
- Ensure all accepted retardant or suppressant mixing and loading procedures are followed.
- Act as liaison between vendor and Air Tanker Base Manager when operating for an established air tanker base.
- Perform SEAT logistical coordination according to the *Interagency Single Engine Airtanker Operations Guide*, PMS 506.
- Coordinate with the local dispatch organization or Air Support Group Supervisor, if assigned to Incident Management Team for mission assignments.

Helibase Manager (HEB1 [6+ helicopters] or HEB2 [1 to 5 helicopters])

Critical Safety Responsibilities

- Obtain briefing from Air Support Group Supervisor.
- Conduct briefings for helibase or helispot personnel and pilots.
- Ensure helibase is set up to accommodate current and planned helicopter operations.
- Ensure helibase air traffic control operations are in effect and coordinate helicopter traffic routes with the Helicopter Coordinator or Air Tactical Group Supervisor.
- Manage appropriate crash-rescue services for the helibase and helispots.

Other Duties

- Report staffing and equipment needs to Air Support Group Supervisor.
- Manage resources and supplies dispatched to helibase.
- Manage helibase retardant mixing and loading: Retardant contract management, logistics planning and coordination, environmental restrictions of mixing operation.
- Display organization and work schedule at each helibase, including helispot organization and assigned radio frequencies.
- Supervise manifesting and loading of personnel and cargo.
- Ensure dust abatement is provided when needed.
- Consider security at each helibase and helispot as appropriate.
- Request special air support items from the Air Support Group Supervisor.
- Maintain agency records and reports of helicopter activities. Coordinate daily cost reports with the Air Support Group Supervisor or Finance Unit.
- Supervise the Helicopter Managers, Deck Coordinator, Mixmaster, and other positions (see Operations Organization Chart).

Deck Coordinator (DECK)

The Deck Coordinator reports to the Helibase Manager or Fixed-Wing Base Manager and is responsible for providing coordination at an aircraft landing area for personnel and cargo movement.

Critical Safety Responsibilities

- Obtain briefing from supervisor.
- Establish emergency landing areas.
- Ensure deck personnel understand crash and rescue procedures.
- Establish and mark landing areas.
- Ensure sufficient personnel are available to safely load and unload personnel and cargo.
- Ensure deck area is properly posted.
- Ensure proper manifesting and load calculations are done.

Other Duties

- Supervise deck management personnel.
- Apply dust abatement when necessary.
- Ensure Air Traffic Control operation is coordinated with the Takeoff and Landing Coordinator.
- Maintain agency records.

Takeoff and Landing Coordinator (TOLC)

The Takeoff and Landing Coordinator reports to the Helibase Manager and is responsible for providing coordination of arriving and departing helicopters and movement around the helibase.

Critical Safety Responsibilities

- Obtain briefing from Helibase Manager.
- Check radio system before commencing operation.
- Coordinate with Radio Operator on helicopter flight routes and patterns.
- Maintain communications with all incoming and outgoing helicopters.

Other Duties

- Coordinate with Deck Coordinator and Parking Tender.

Aircraft Base Radio Operator (ABRO)

The Aircraft Base Radio Operator reports to the Helibase Manager or Fixed-Wing Base Manager and is responsible for establishing communication between incident assigned aircraft and airbases, Air Tactical Group Supervisor, Air Operations Branch Director, and the Takeoff and Landing Coordinator.

Critical Safety Responsibilities

- Obtain briefing from Base Manager.
- Maintain a log of all aircraft takeoffs and landings, estimated times of arrival (ETAs), estimated times of departure (ETDs), and flight route check-ins.
- Establish and enforce proper radio procedures.
- Immediately notify supervisor of any overdue or missing aircraft.
- Understand crash and rescue procedures.
- Establish radio communication with the aerial supervisor.

Other Duties

- Obtain Air Operations Summary Worksheet (ICS 220).
- Notify Takeoff and Landing Coordinator of incoming aircraft.
- Verify daily radio frequencies with Base Manager.

Helicopter Crewmember (HECM)

The Helicopter Crewmember reports to the Helicopter Manager and is part of a Helicopter Module. The HECM is primarily responsible for supporting ground-based operations of the helicopter he or she is assigned to.

Critical Safety Responsibilities

- Obtain briefing from the Helicopter Manager.

- Know and understand responsibilities of the Parking Tender (PARK) and Loadmaster (LOAD) positions.

- Perform Helispot Manager (HESM), Deck Coordinator (DECK), Aircraft Base Radio Operator (ABRO), Takeoff and Landing Coordinator (TOLC) duties if qualified to NWCG PMS 310-1 standards.

Air Tactical Group Supervisor (ATGS)

The Air Tactical Group Supervisor (ATGS) reports to the Air Operations Branch Director on Type 1 and Type 2 incidents and to the Incident Commander or Operations on initial attack or Type 3 incidents. The ATGS is responsible for managing the incident airspace and coordinating the fixed- and rotary-wing aircraft operations over an incident. (See NWCG *Interagency Aerial Supervision Guide*, PMS 505. **Note**: Aerial Supervision Modules (ASM) can function as an ATGS or an ATCO.)

Critical Safety Responsibilities

- Ensure that the aerial supervision pre-mission responsibilities are completed (pilot/aircraft qualification, flight and duty, and maintenance schedules known).

- Provide the pilot with a mission briefing, and prepare for the flight (ATGS and the pilot).

- En route procedures: Establish flight following.

- Airspace coordination procedures.

- Entering incident airspace FTA.

- Incoming aircraft: Conduct initial briefing, tactical briefing, and departure briefing.

- Perform Air traffic coordination: horizontal and vertical separation, holding areas and initial points (IP), and entrance and exit corridors.

- Determine that adequate VHF-FM (used for air-to-ground) and VHF-AM (used for air-to-air) radio frequencies are used or requested. Establish contact with Incident Commander, Operations, DIVS, Helibase, ATCO, and HLCO.

- Incident flight following: Establish procedures with dispatch, air operations, and other aircraft.

- Identify aviation safety issues, and mitigate any hazards.

- Determine the need for a TFR and the dimensions of the TFR.

- Coordinate approved flights of nonincident aircraft in TFR.
- Assign air resources according to Operations' or Incident Commander's strategy, tactics, and mission priorities.
- Coordinate with ground forces.
- Provide fire information and sizeup for strategic and tactical planning; make recommendations based on incident objectives and observed situation.
- Provide safety oversight to ground crews: Drop zone clearance, and adverse weather and fire behavior. (Note: The aerial supervisor can only advise about potential safety zones and escape routes; ground forces must verify.)
- Coordinate between types of aerial supervisors: Workload management, briefings, and maintaining aerial supervision continuity.
- Institute emergency procedures.
- Oversee in-flight emergencies.
- Oversee missing aircraft and aircraft mishap.
- Coordinate medevac of incident personnel.

Other Duties

- Provide initial response sizeup information for Dispatch and responding resources.
- Inform AOBD of tactical recommendations affecting the air operations portion of the Incident Action Plan.
- Report to Air Operations or Unit Aviation Manager on incidents or accidents.
- Determine the procedures for ordering tactical aerial resources.
- Complete the appropriate aircraft contract or fleet utilization records.
- Maintain coordination with air bases supporting the incident.
- Inform AOBD of special aircraft and/or pilot restrictions.
- Ensure compliance with each agency's operations checklist for day and night operations.
- Use standard target description.

Air Tanker/Fixed Wing Coordinator (ATCO)

The Air Tanker/Fixed Wing Coordinator reports to the Air Tactical Group Supervisor and is responsible for coordinating assigned air tanker operations at the incident. The coordinator is always airborne.

Critical Safety Responsibilities

- Obtain briefing from the Air Tactical Group Supervisor.
- Determine all aircraft including air tankers and helicopters operating within incident area of assignment.
- Survey incident area to determine situation, aircraft hazards, and other potential problems.
- Coordinate the use of assigned ground-to-air and air-to-air communications frequencies.
- Ensure air tanker flight crews know appropriate operating frequencies.
- Determine incident air tanker capabilities and limitations for specific assignments.
- Coordinate with Air Tactical Group Supervisor and assign geographical areas for air tanker operations.
- Implement air safety procedures. Immediately correct unsafe practices or conditions.

Other Duties

- Receive assignments, assign missions, schedule flights, and supervise air tanker activities.
- Provide information to ground resources.
- Inform Air Tactical Group Supervisor of overall incident conditions including aircraft malfunction or maintenance difficulties.
- Inform Air Tactical Group Supervisor when mission is completed and reassign air tankers as directed.
- Report incidents or accidents.
- Maintain records of activities.

Helicopter Coordinator (HLCO)

The Helicopter Coordinator reports to the Air Tactical Group Supervisor and is responsible for coordinating tactical or logistical helicopter mission(s) at the incident.

Critical Safety Responsibilities

- Obtain briefing from the Air Tactical Group Supervisor.
- Survey assigned incident area to determine situation, aircraft hazards, and other potential problems.
- Coordinate with Air Tactical Group Supervisor in establishing locations and takeoff and landing patterns for helibase(s) and helispot(s).
- Coordinate the use of assigned ground-to-air and air-to-air communications frequencies with the Air Tactical Group Supervisor.
- Ensure that all assigned helicopters know appropriate operating frequencies.
- Coordinate geographical areas for helicopter operations with Air Tactical Group Supervisor, and make assignments.
- Implement air safety procedures. Immediately correct unsafe practices or conditions.

Other Duties

- Ensure that approved night flying procedures are in operation.
- Coordinate activities with Air Tactical Group Supervisor, Air Tanker Coordinator, Air Support Group Supervisor, and ground personnel.
- Inform Air Tactical Group Supervisor when mission is completed, and reassign helicopter as directed.
- Report incidents or accidents.
- Maintain records of activities.

Helispot Location and Construction

A helispot is a natural or improved takeoff and landing area intended for temporary or occasional helicopter use. It may or may not have road access.

Points to consider in locating and constructing helispots are (see *IRPG* for more detailed information):

- Allow for takeoff and landing from all directions into the wind as often as possible.
- Situations that require a maximum-power takeoff and landing have a higher risk than those that allow for forward flight during takeoff and landing, and payloads will be reduced.
- When using roads or turnouts, ensure adequate traffic control. On dirt roads and dozer lines, visibility can be severely impaired due to dust.

Principles of Retardant Application

- Determine tactics direct or indirect based on fire sizeup and resources available.
- Establish an anchor point and work from it. Use direct attack only when ground support is available or extinguishment is feasible.
- Plan drops so they can be extended or intersected effectively.
- Maintain honest evaluation and effective communication between the ground and air.
- Monitor retardant effectiveness, and adjust its use accordingly.
- Refer to the *IRPG* for additional information.

PLANNING

Organization Chart

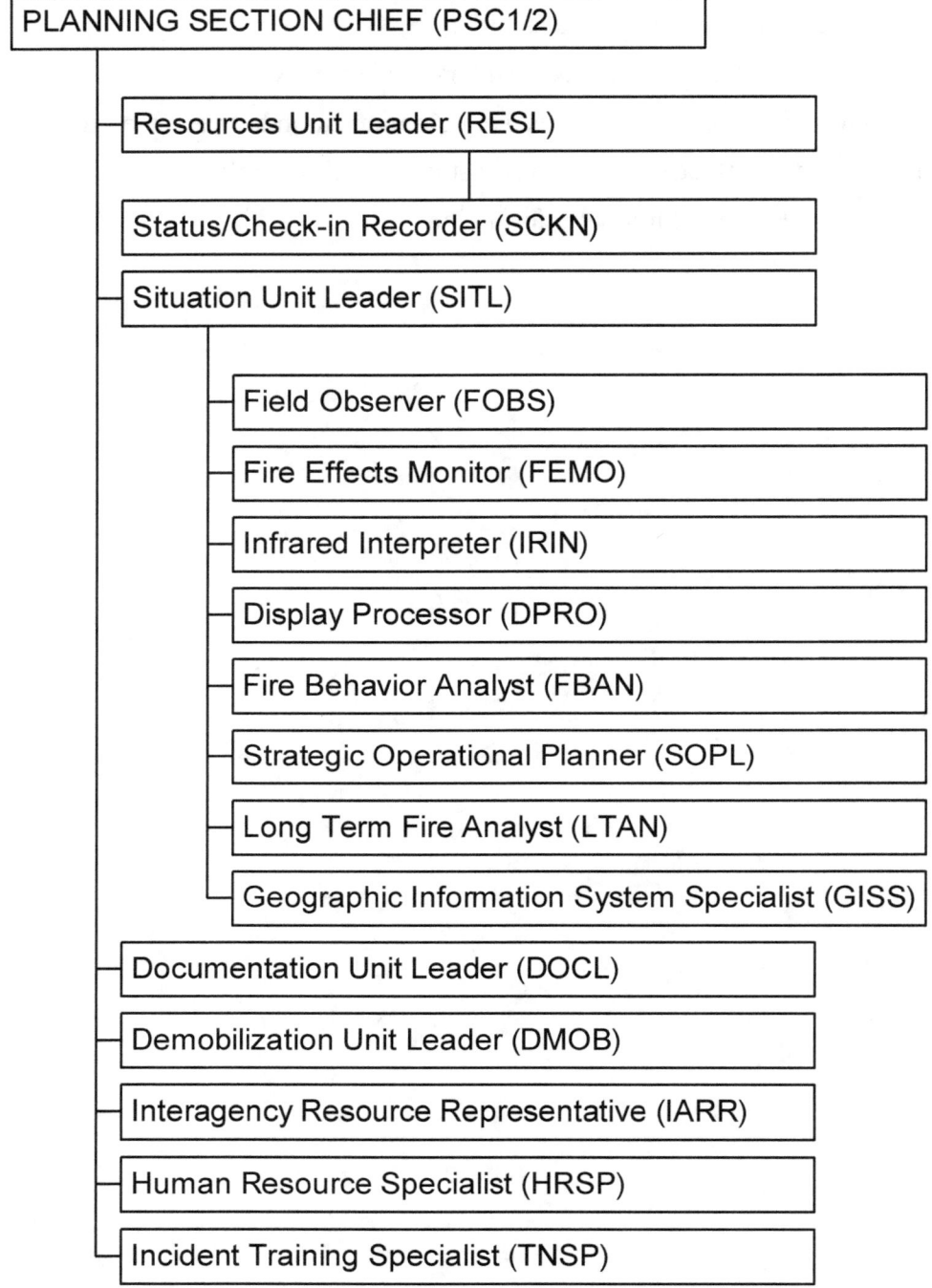

Position Checklists

Planning Section Chief (PSC1/2)

The Planning Section Chief, a member of the General Staff, is responsible for collecting, evaluating, disseminating, and using information about the development of the incident, status of resources, and demobilization of the incident.

Information is needed to understand the current situation, predict probable course of incident events, prepare alternative strategies and control operations for the incident, and provide for an orderly and economical demobilization of the incident.

Critical Safety Responsibilities

- Conduct Planning Meetings and operational briefings.
- Supervise preparation of IAP (see Planning Process), and ensure sufficient copies are available for distribution through Unit Leader level.
- Advise General Staff of any significant changes in incident status.
- Prepare and distribute Incident Commander's orders.
- Ensure that information concerning special environmental protection needed is included in the IAP.
- Establish information requirements and reporting schedules for all ICS Organizational elements for use in preparing the IAP.
- Instruct Planning Section Units in distribution of information.

Other Duties

- Assemble information on alternative strategies.
- Perform operational planning for Planning Section.
- Ensure that normal agency information collection and reporting requirements are met.
- Prepare recommendations for release of resources (for approval by the Incident Commander).
- Ensure demobilization plan and schedule are developed and coordinated with Command, General Staff, and Agency Dispatchers.
- Establish a communications link between the agency Demobilization Organization and the incident Demobilization Unit.

Resources Unit Leader (RESL)

The Resources Unit Leader is responsible for establishing all incident check-in activities; preparing and processing resource status information; preparing and maintaining displays, charts, and lists that reflect the current status and location of suppression resources, transportation, and support vehicles; and maintaining a master check-in list of resources assigned to the incident.

Critical Safety Responsibilities

- Gather, post, and maintain current incident resource status. including transportation, support vehicles, and personnel.
- Maintain master list of all resources checked in at the incident.
- Prepare Organization Assignment List (ICS 203) and Incident Organization Chart (ICS 207).
- Assemble and disassemble Task Forces/Strike Teams as requested by Operations.
- Participates in the Tactics Meeting and completes the ISC 215 with the Operations Section Chief.
- Prepare Division Assignment List(s) (ICS 204) after the Planning Meeting.

Other Duties

- Establish check-in function at incident locations.
- Verify that all resources are checked in.
- Using the Incident Briefing (ICS 201), prepare and maintain the Command Post display (organization chart and resource allocation and deployment sections of display).
- Establish contacts with incident facilities and maintain resource status information.
- Participate in Planning Meetings as required by the Planning Section Chief.
- Provide resource summary information to Situation Unit as requested.
- Continually identify resources that are surplus to the suppression needs.

Status/Check-in Recorder (SCKN)

Status/Check-in Recorders are used at each check in location to ensure that all resources assigned to an incident are accounted for. (Where practical, employ Demobilization Unit Leader as a Status/Check-in Recorder to ensure complete information is obtained at check-in.)

Critical Safety Responsibilities

- Transmit check-in information to Resources Unit on a regular, prearranged schedule.
- Forward completed Incident Check-in Lists (ICS 211) to the Resources Unit.
- Prepare, post, and maintain Resource Status Cards (ICS 219).

Other Duties

- Establish communications with the Communication Center.
- Post signs so arriving resources can easily find the check-in locations.

Situation Unit Leader (SITL)

The Situation Unit Leader is responsible for collecting and organizing incident status and information and evaluating, analyzing, and displaying that information for use by ICS personnel and agency Dispatchers.

Critical Safety Responsibilities

- Collect and analyze situation data.
- Obtain available preattack plans, mobilization plans, maps, and photographs.
- Prepare predictions at periodic intervals or upon request of the Planning Section Chief.
- Prepare the Incident Status Summary (ICS 209).

Other Duties

- Obtain and analyze infrared data as applicable.
- Post data on Unit work displays and Command Post displays at scheduled intervals.
- Participate in Planning Meetings as required by the Planning Section Chief.
- Provide information on transportation system to Ground Support Unit Leader for the Transportation Plan.
- Provide photographic services and maps.
- Maintain Situation Unit records.
- Maintain incident history on maps and narrative from Initial Attack to final demobilization.

Field Observer (FOBS)

The Field Observer is responsible for collecting incident status information from personal observations at the incident and providing this information to the Situation Unit Leader, Division Supervisor, and other fireline resources as directed. The information may include but is not limited to, fire perimeter location, onsite weather, fire behavior, fuel conditions, and fire effects information needed to assess firefighter safety and whether the fire is achieving established incident objectives and requirements.

Critical Safety Responsibilities

- Monitor, obtain, and record weather data.
- Monitor and record fire behavior data.
- Recognize and report atmospheric characteristics that influence fire behavior.

- Monitor and record smoke management information.
- Recon the fire area assigned.
- Plot fire perimeter on a map.
- Assist in preparing maps for use in Situation Unit, Incident Command Post, and IAP to ensure accuracy.
- Immediately report any condition that may cause danger or be a safety hazard to personnel.
- Let appropriate Operations overhead know you are in the area.

Other Duties

- Determine location of assignment, types of information required, priorities, time limits for completion, methods of communication and documentation, and method of transportation.
- Observe and record first order fire effects.
- Provide completed observation logs or forms for weather, fire behavior, fuel conditions, and fire effects as assigned. Summarize observations as requested by supervisor.
- Attend end-of-shift debriefings of operations personnel, and at other times as appropriate, to obtain situation information.
- Coordinate an efficient transfer of position duties when mobilizing and demobilizing (e.g., IMT or host agency).

Fire Effects Monitor (FEMO)

The Fire Effects Monitor is responsible for collecting incident status information from personal observations at the incident, and providing this information to the module leader, crew boss and other fireline supervisor as directed. The information may include but is not limited to fire perimeter location, onsite weather, fire behavior, fuel conditions, smoke, and fire effects information needed to assess firefighter safety and whether the fire is achieving established incident objectives and requirements.

Critical Safety Responsibilities

- Monitor, obtain, and record weather data.
- Monitor and record fire behavior data.
- Recognize and report atmospheric characteristics that influence fire behavior.
- Monitor and record smoke dispersion and air quality information.
- Recon the fire area assigned.
- Plot fire perimeter on a map.
- Assist in preparing maps for use in Situation Unit, Command Post, and IAP to ensure accuracy.

- Immediately report any condition that may cause danger or be a safety hazard to personnel.
- Maintain communication with immediate supervisor while moving around the incident area.

The prerequisite for the position of Fire Effects Monitor is Firefighter Type 2, and the ordering or use of this position as an independent single resource on a wildfire or high complexity prescribed fire is prohibited. A Fire Effects Monitor may be used on wildfires or high complexity prescribed fires as a member of a crew or module under the direct supervision of a Single Resource Boss.

Fire Effects Monitors with a higher level of fireline qualification may be used as appropriate to their additional qualifications (e.g., a Fire Effects Monitor who is also Single Resource Boss-qualified may be given a fireline assignment as an independent resource based on that Single Resource Boss qualification).

Other Duties

- Determine location of assignment, types of information required, priorities, time limits for completion, methods of communication and documentation, and method of transportation.
- Observe and record first-order fire effects.
- Provide completed observation logs or forms for weather, fire behavior, fuel conditions, and fire effects as assigned. Summarize observations as requested by supervisor.

Infrared Interpreter (IRIN)

The Infrared Interpreter directs infrared mapping operations when assigned.

- Interpret imagery and plot findings on aerial photos or maps.
- Arrange for missions with infrared aircraft crew liaison, including objectives of flight, timing, areas needing particular attention, and imagery delivery.
- Keep abreast of aircraft or crew limitations.
- Keep Planning Section currently advised of findings.

Display Processor (DPRO)

The display processor is responsible for the display of incident status information obtained from field observers, aerial and ortho photographs, and infrared data.

Critical Safety Responsibilities

- Assist Situation Unit Leader in analyzing and evaluating field reports.
- Develop required displays in accordance with time limits for completion.
- Support special requirements for development of incident maps.

Other Duties

- Determine:
    - ✓ Location of work assignments.
    - ✓ Numbers, types, and locations of displays required.
    - ✓ Priorities.
    - ✓ Map requirements for IAPs.
    - ✓ Time limits for completion.
    - ✓ Field observer assignments and communications means.
- Demobilize incident displays in accordance with incident demobilization plan.

Fire Behavior Analyst (FBAN)

The Fire Behavior Analyst is responsible for collecting weather data, developing strategic and tactical fire behavior information, predicting fire growth, and interpreting fire characteristics for use by incident overhead.

Critical Safety Responsibilities

- Manage weather data collection system and Incident Meteorologist (IMET) and Weather Observers. (These positions are not in the *Wildland Fire Qualification System Guide*, PMS 310-1, June 2012.)
- Provide weather information and other pertinent information to Situation Unit Leader for inclusion in Incident Status Summary Report (ICS 209).
- Develop tactical fire behavior information in support of the IAP.
- Prepare a written fire behavior forecast that includes safety considerations for each operational period.
- Participate in operational briefings to present fire behavior predictions and to answer questions related to fire behavior, interpretations, and safety.
- Monitor actual fire behavior to validate predictions, document behavior, and anticipate potential safety problems.
- Ensure all affected incident personnel are advised of anticipated changes in weather conditions or predictions.
- Provide site-specific fire behavior predictions, as requested.

Other Duties

- Participate in Planning Meetings as directed by the Situation Unit Leader.
- Collect, review, and compile fire history, fuel data, and information about topography and fire barriers.

Strategic Operational Planner (SOPL)

The SOPL position is responsible for developing courses of action on long-term wildfire events. The courses of action for these wildfires may include both protection and resource benefit objectives. The SOPL may be ordered by and work for the host unit, the Geographic Area Coordination Center (GACC), or the IMT assigned to the fire.

Critical Safety Responsibilities

- Develop a course of action for long term fire events, in collaboration with the local affected agencies and incident management organization.
- Obtain long-term assessments of weather and fire behavior to develop strategies and tactics associated with the course of action.
- Use risk assessment information in developing the course of action. Evaluate current course of action to ensure that resource and protection objectives can be met through plan implementation.

Other Duties

- Review and understand agency procedures and policies
- Assess the situation to understand support needs.
- Understand the area's fire management plans and objectives for managing wildfires within the impacted fire management units.
- Review strategic objectives and management requirements as stated in the wildland fire decision document.
- Participate in daily briefings and Planning Meetings to review recommended actions and ensure they are consistent with agency direction and long-range plans. Provide possible alternatives and contingency actions, and ensure values will be protected by proposed management actions.
- Coordinate with local resource advisors.
- Evaluate consistency between strategic objectives and management requirements with incident objectives and incident requirements as stated in the wildland fire decision document.
- Recommend modifications to the proposed course of action in response to changing conditions and forecasts.
- Develop cost estimates for the proposed management action points.
- Participate in AARs and closeouts with local Agency Administrators.
- Develop appropriate material for transfer of command as required.

Long Term Fire Analyst (LTAN)

Provides probabilistic and deterministic information on long-term fire advancement, fire behavior, and spread direction, based on local information, topography, historic and current fire spread, and with historic and current fire weather data.

Critical Safety Responsibilities

- Manage weather data collection system and Incident Meteorologists and Weather Observers.
- Access and analyze historic weather records from local stations to determine important thresholds and season-ending criteria.
- Produce products and provide support for decision making and planning through the use of long- and short-term models, such as RERAP (Rare Event Risk Assessment Process), FSPro (Fire Spread Probability), FARSITE, and FLAMMAP.
- Assess and document fire growth toward multiple points of interest.
- Provide short-, medium- and long-term fire growth estimates and projections.
- Assist with developing management action points in relation to expected fire behavior and time.
- Characterize risk of potential strategy decisions, and provide feedback.

Other Duties

- Collect, review, and compile fire history, fuel data, and information about topography and fire barriers.
- Provide input to IAPs as requested, and review plans for consistency with fire behavior modeling outputs.
- Obtain or provide smoke management predictions as requested.
- Provide clear documentation of assumptions and changes used when developing fire spread predictions.

Geographic Information System Specialist (GISS)

The Geographic Information System Specialist is responsible for providing timely and accurate spatial information to the Situation Unit Leader about the incident to be used by all facets of the IMT.

Critical Safety Responsibilities

- Prepare incident maps and displays as requested by the Situation Unit Leader.
- Provide updated maps as required for IAPs.

Other Duties

- Understand expectations and attendance required at Planning Meetings.
- Participate in functional area briefings and AARs.
- Provide written documentation, digital data, and products developed during the incident to the Documentation Unit.
- Complete digital analysis as requested.
- Ensure metadata is updated and maintained.
- Download or use spatial data provided by local agencies.
- Follow NWCG Geographic Information System (GIS) standard operating procedures.

Documentation Unit Leader (DOCL)

The Documentation Unit Leader is responsible for maintaining accurate and complete incident files, providing duplication services to incident personnel, and packing and storing incident files.

- Establish and organize incident files.
- Establish duplication service and respond to requests.
- Retain and file duplicate copies of official forms and reports, including those generated by computers.
- Check on accuracy and completeness of records.
- Provide duplicates of forms and reports.
- Prepare incident documentation when requested.
- Maintain, retain, and store incident files.

Demobilization Unit Leader (DMOB)

The Demobilization Unit Leader is responsible for preparing the Demobilization Plan and schedule. The Demobilization Unit Leader assists the Command and General Staff in ensuring an orderly, safe, and efficient movement of personnel and equipment from the incident.

- Review and continually monitor incident resource records (Incident Briefing Form (ICS 201), Incident Check-In List (ICS 211), Resource Status Cards (ICS 219), and IAP) to determine probable size of demobilization effort.
- Obtain Incident Commander's demobilization objectives and priorities.
- Meet with Agency Representatives to determine:
  - ✓ Personnel rest, hygiene, and safety needs.
  - ✓ Coordination procedures with agencies.
  - ✓ Local and national demobilization priorities.
- Be aware of ongoing Operations Section resource needs.
- Obtain identification and description of surplus resources and probable release times.
- Determine finance, supply, and other incident checkout stops.
- Establish and post check out procedures.
- Determine incident logistics and transportation capabilities needed to support the demobilization effort.
- Establish communications with appropriate off-incident facilities.
- Get approval of Demobilization Plan (Incident Commander, Planning Section Chief, agency, etc.).
- Distribute Plan and any amendments.
- Monitor and supervise implementation of Demobilization Plan.

Interagency Resource Representative (IARR)

The Interagency Resource Representative may be assigned to an incident to serve as the sending area's representative for crews, overhead, and equipment assigned to an incident. The Interagency Resource Representative is responsible to the home unit to coordinate, through the IMT, the well-being of all resources assigned from the home unit. This position will normally check in with the Planning Section but is not an incident resource.

- Secure and maintain a complete list of names, home agencies and units, Social Security numbers, etc., of all personnel assigned to the incident from the sending area. Verify and update list(s) as needed at the incident.
- Establish contact with the IMT to provide information and assistance to the team during resource check-in and initial assignment.
- Coordinate activities with appropriate Agency Representatives.

- Establish a work location. Advise the team and assigned resources about that location.
- Whenever feasible, maintain daily contact with a representative of each appropriate resource.
- Provide assistance to appropriate personnel on timekeeping, commissary, travel, accidents, injuries, personnel problems or emergencies, and other administrative needs.
- Maintain daily contact with the sending area to exchange information about the status of resources.
- Assist in resolving disciplinary cases as requested by the team or the sending area.
- Provide input as to the use of assigned resources.
- Assist the team in providing for the well-being and safety of assigned resources.
- Assist the team in determining the need for and preparation of special reports or documents.
- Assist the team in investigating accidents involving assigned personnel.
- Maintain contact with assigned personnel that have been hospitalized or otherwise separated from their Unit.
- Assist the team in completing all required forms, reports, and documentation before assigned resources depart from the incident.
- Assist the team in demobilizing assigned resources.
- Provide the sending Unit with pertinent paperwork and evaluations relating to the resources for which they are responsible.

Human Resource Specialist (HRSP)

The Human Resource Specialist is responsible for monitoring civil rights and related human resource activities to ensure that appropriate practices are followed. Work is normally conducted in a Base Camp environment but may involve tours of the fireline, other Camps, and rest and recuperation (R&R) facilities.

- Establish contact with the Planning Section Chief to determine placement within the organization.
- Provide a point of contact for incident personnel to discuss civil rights and human resource concerns.
- Participate in daily briefings and Planning Meetings to provide appropriate civil rights and human resource information.
- Prepare civil rights messages for inclusion in IAP(s).
- Post-civil rights or other human resource information on bulletin boards and other appropriate Message Centers.
- Monitor whether a positive working environment, supportive of cultural diversity, is maintained and enhanced for all personnel.

- Conduct awareness sessions as needed. Use civil rights or human resource videotapes when appropriate.
- Establish and maintain effective work relationships with agency representatives, liaisons, and other personnel in the Incident Command.
- Refer concerns about pay, food, sleeping areas, transportation, and shift changes to the appropriate incident staff, taking into account civil rights and human resource factors.
- Receive and verify reports of inappropriate behavior that occur on the incident.
- Take steps to correct inappropriate acts or conditions through appropriate lines of authority.
- Give high priority to informally resolving issues before the individuals leave the incident.
- Provide referral information if a complaint cannot be resolved during the incident.
- Conduct followup, as needed, depending upon the seriousness of the infraction.
- Prepare and submit reports and related documents.

Incident Training Specialist (TNSP)

An Incident Training Specialist may help achieve and oversee training opportunities on an incident. To be effective, training activities must be coordinated at all levels.

- Identify training opportunities on the incident.
- Review trainee assignments and modify, if appropriate.
- Inform Resources Unit of trainee assignments.
- Brief trainees and trainers on training assignments and objectives.
- Make followup contacts on the job to provide assistance and advice for trainees to meet training objectives.
- Ensure trainees receive their performance evaluation and complete position task book as assigned.
- Prepare formal report for trainees' home units.

Planning Process

The *ERFOG* contains information appropriate for use in almost any incident situation. Not all incidents require written plans. The need for written plans and attachments is based on incident requirements and the decision of the Incident Commander.

The Operations Section Chief should have a draft Operational Planning Worksheet (ICS 215) completed before the Planning Meeting. In addition, an Incident Action Plan Safety Analysis (ICS 215A) must be completed for each Planning Meeting. The form should be completed as a draft before the meeting and discussed as part of the planning process.

Incident objectives and strategy should be established before the Planning Meeting. For this purpose, it may be necessary to hold a Strategy Meeting before the Planning Meeting.

The planning process works best when the incident perimeter and proposed control lines are divided into logical geographical units. The tactics and resources are then determined for each of the Planning Units. Finally, the Planning Units are combined into segments or Divisions, using span-of-control guidelines.

Demobilization

The Incident Commander is responsible to the host agency for demobilization. Demobilization is an important part of total incident management and requires the attention of the Incident Commander and the Command and General Staff.

The Planning Section Chief must establish an adequate demobilization organization, in a timely fashion, to provide for an orderly and economic demobilization of the incident. Using the Demobilization Unit Leader as a Status/Check-in Recorder early in the incident, where possible, facilitates the collection of resource information necessary to develop a Demobilization Plan. The complexity of the incident, kinds and types of resources, and the level of resources involved (local, regional, or national) dictate the size and expertise needed by the demobilization organization. Resources must be released, returned to their home units, rested, and rehabilitated as soon as possible so they will be ready for their next assignment.

The Demobilization Unit Leader must obtain input from a number of others to develop a complete Plan. The IC and General Staff need to provide input and totally support the Plan. The agency dispatcher must provide input from all coordination levels. If Area Command has been established, they should provide their input directly to the incident.

LOGISTICS

Organization Chart

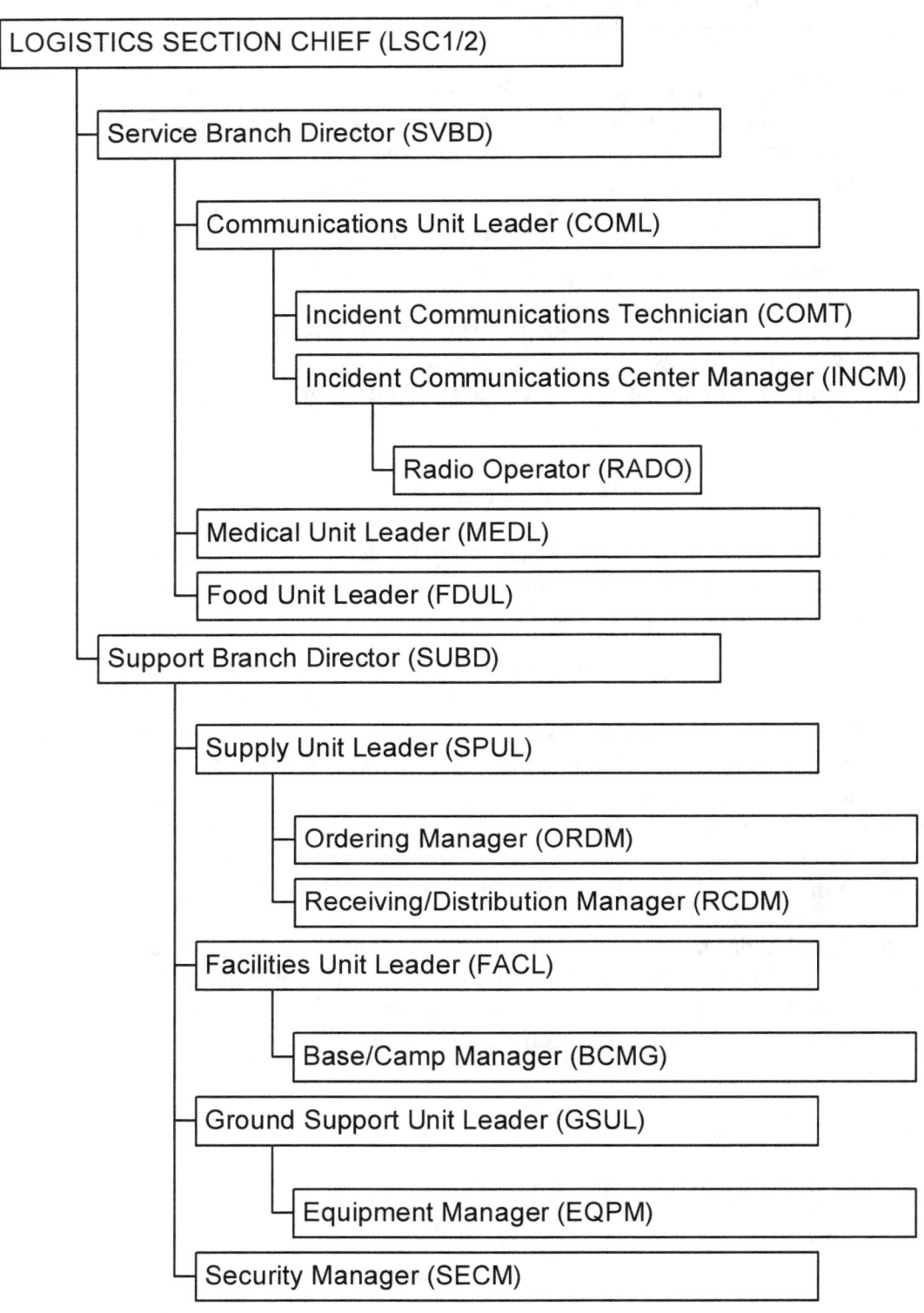

Position Checklists

Logistics Section Chief (LSC1/2)

The Logistics Section Chief, a member of the General Staff, is responsible for providing facilities, services, and material in support of the incident. The Logistics Section Chief participates in developing and implementing the IAP and activates and supervises the Branches and Units within the Logistics Section.

Critical Safety Responsibilities

- Obtain briefing from Agency Administrator and/or outgoing Incident Commander, and gather intelligence.
- Before your arrival, collect information from outgoing Logistics personnel responsible for the incident.
- Identify service and support requirements for planned and expected operations.
- Participate in preparing the IAP.
- Ensure Communication Plan, Medical Plan, and Transportation Plan are updated and provided to Planning Section.

Other Duties

- Advise on current service and support capabilities.
- Update Incident Commander on accomplishments and/or problems.
- Consider demobilization before the actual need to release excess section resources.

Service Branch Director (SVBD)

The Service Branch Director is responsible for managing all service activities at the incident. The Service Branch Director supervises the operations of the Communications, Medical, and Food Unit Leaders.

Critical Safety Responsibilities

- Supervise Service Branch Leaders.
- Ensure Communications and Medical Plans are updated and provided to the Planning Section.

Other Duties

- Advise on current service capabilities.
- Inform Logistics Section Chief of Branch activities.
- Update Logistics Section Chief on accomplishments and problems.
- Consider demobilization before the actual need to release excess Branch resources.

Communications Unit Leader (COML)

The Communications Unit Leader, under the direction of the Service Branch Director or Logistics Section Chief, is responsible for developing plans for the effective use of incident communications equipment and facilities; installing and testing communications equipment; supervising the Incident Communications Center; distributing communications equipment to incident personnel; and maintaining and repairing communications equipment.

Critical Safety Responsibilities

- Establish adequate communications for the incident.
- Advise Operations Section on communications capabilities and limitations.
- Provide technical information, as required, on limitations and adequacy of communications systems in use, equipment capabilities, equipment available, and potential problems.
- Develop the daily Incident Radio Communications Plan (ICS 205).

Other Duties

- Establish the Communications Unit and Message Centers.
- Establish an equipment accountability system.
- Maintain records on communications equipment.
- Determine location of repeaters.
- Determine what communication networks are established or need to be established.

Incident Communications Technician (COMT)

The Incident Communications Technician works under the direction of the Communication Unit Leader and is responsible for installing, maintaining, and tracking communications equipment.

Critical Safety Responsibilities

- Assist in designing communications system for incident to meet operational needs.
- Install and test communications equipment.
- Clone or program radios.
- Repair and/or replace communications equipment.

Other Duties

- Issue and track communications equipment.
- Identify operational restrictions.

Incident Communications Center Manager (INCM)

The Incident Communications Center Manager is responsible for receiving and transmitting radio and telephone messages among and between personnel and to provide dispatch services at the incident.

Critical Safety Responsibilities

- Establish communications procedures.
- Determine frequencies in use.
- Coordinate with local dispatch center to ensure information is transmitted.
- Keep track of critical cell phone, telephone, and satellite numbers.

Other Duties

- Set up Communications Center.
- Check out equipment to firefighters.
- Receive and transmit messages internally and externally.
- Maintain a record of unusual incident occurrences.

Radio Operator (RADO)

The Radio Operator works in the Communications Unit, under the Logistics Section. The immediate supervisor for the Radio Operator is the Incident Communications Center Manager who manages the Incident Communications Center (ICC). In the absence of an Incident Communications Center Manager, the Communications Unit Leader will supervise the Radio Operator position. The Communications Technician also works in the Communications Unit. Often the Communications Technician requests assistance from the Radio Operator to help clone and check out radios.

Critical Safety Responsibilities

- Effectively communicate information to incident personnel following proper radio and/or telephone procedures.
- Acknowledge requests and provide feedback.
- Use appropriate communication protocol when responding to emergency situations.

Other Duties

- Document all calls and radio transmissions.
- Correctly fill out and process appropriate forms (General Message (ICS 213), Status Change Card (ICS 210), Radio Logs, Telephone Logs).
- Coordinate the efficient transfer of position duties when mobilizing to and demobilizing from the incident.

Medical Unit Leader (MEDL)

The Medical Unit Leader is primarily responsible for developing the Medical Plan, obtaining medical aid and transportation for injured or ill incident personnel, and preparing reports and records.

The Medical Unit may also assist Operations in supplying medical care and assistance to civilian casualties at the incident.

Critical Safety Responsibilities

- Determine level of emergency medical activities performed before activating Medical Unit.
- Prepare the Medical Plan (ICS 206).
- Prepare procedures for major medical emergencies.
- Declare major medical emergency(s) as appropriate.
- Provide medical aid, supplies, and transportation.
- Audit use of "over-the-counter" medications being dispensed by the Medical Unit to discourage improper use or abuse.

Other Duties

- Prepare medical reports.
- Contact Compensation-For-Injury Specialist to establish coordination procedures.
- Provide space for Compensation-For-Injury Specialist as needed.

Food Unit Leader (FDUL)

The Food Unit Leader is responsible for determining feeding requirements at all incident facilities and for menu planning, determining cooking facilities required, food preparation, serving, providing potable water, and general maintenance of the food service areas.

Critical Safety Responsibilities

- Obtain necessary equipment and supplies to operate food service facilities at Base and Camps.
- Provide sufficient potable water to meet food service needs.
- Ensure appropriate health and safety measures are taken.
- Keep inventory of food on hand, and check in food orders.

Support Branch Director (SUBD)

The Support Branch Director is responsible for developing and implementing logistics plans in support of the IAP. The Support Branch Director supervises the operations of the Supply, Facilities, and Ground Support Units.

Critical Safety Responsibilities

- Determine level of service needed to support operations.

Supply Unit Leader (SPUL)

The Supply Unit Leader is responsible for ordering personnel, equipment, and supplies; receiving and storing all supplies for the incident; maintaining an inventory of supplies; and servicing nonexpendable supplies and equipment.

Critical Safety Responsibilities:

- Develop and implement safety and security requirements.

Other Duties

- Arrange for receiving ordered supplies.
- Order, receive, store, and distribute supplies and equipment.
- Order personnel, supplies, and equipment as requested.
- Maintain inventory and accountability of supplies and equipment.
- Responsible for proper disposal of expendable supplies and hazardous wastes.

Ordering Manager (ORDM)

The Ordering Manager is responsible for placing all orders for supplies and equipment for the incident.

- Establish ordering procedures.
- Identify incident personnel who have ordering authority.
- Verify what has already been ordered.
- Ensure order forms are filled out correctly.
- Place orders in a timely manner.
- Consolidate orders when possible.
- Keep Receiving/Distribution Manager informed of orders placed.
- Resolve ordering problems as they occur.

Receiving/Distribution Manager (RCDM)

The Receiving/Distribution Manager is responsible for receiving and distributing all supplies and equipment (other than primary resources) and the service and repair of tools and equipment.

Critical Safety Responsibilities

- Develop security needs for the supply area.

Other Duties

- Establish procedures for operating the supply area.
- Set up appropriate record system.
- Maintain inventory of supplies and equipment.

Facilities Unit Leader (FACL)

The Facilities Unit Leader is responsible for laying out and operating incident facilities (Base, Camp(s), and ICP) and managing Base and Camp(s) operations. Each Base and Camp may be assigned a manager.

Critical Safety Responsibilities

- Provide facility maintenance services: sanitation, lighting, clean up, and potable water.

Other Duties

- Participate in Logistics Section/Support Branch planning.
- Determine requirements for each established facility.
- Prepare layouts of incident facilities.
- Provide Base and Camp Managers.
- Provide sleeping facilities.

Base/Camp Manager (BCMG)

The Base/Camp Manager is responsible for appropriate sanitation and facility management services in the assigned Base or Camp.

Critical Safety Responsibilities

- Ensure compliance with all applicable safety regulations.
- Determine or establish special requirements or restrictions on facilities or operations.
- Ensure that all facilities and equipment are set up and functioning properly.
- Supervise the setup of sleeping, shower, and sanitation facilities.

Other Duties

- Provide all necessary facility maintenance services.

Ground Support Unit Leader (GSUL)

The Ground Support Unit Leader is responsible for transporting personnel, supplies, food, and equipment; fueling, servicing, maintaining, and repairing vehicles and other ground support equipment; supporting out-of-service resources; and developing and implementing the Incident Transportation Plan.

Critical Safety Responsibilities

- Prepare a Transportation Plan for approval by the Logistics Section Chief (obtain traffic data from the Planning Section).
- Mark and correct road system safety hazards, and maintain incident roads.
- Ensure driver familiarity with conditions. Coordinate with Safety Officer and Agency Representatives.
- Conduct incident road system survey to determine traffic management and maintenance requirements.
- Arrange for, activate, and document fueling, maintenance, and repair of ground resources.

Other Duties

- Post signs at drop points, water sources, road junctions, etc.

Equipment Manager (EQPM)

The Equipment Manager provides service, repair, and fuel for all apparatus and equipment; provides transportation and support vehicle services; and maintains records of equipment use and service provided.

Critical Safety Responsibilities

- Inspect equipment condition and ensure equipment is covered by an appropriate agreement.
- Coordinate with Agency Representatives on service and repair as required.
- Determine supplies (gasoline, diesel, oil, and parts) needed to maintain equipment in efficient operating condition.

Other Duties

- Provide transportation and support vehicles.
- Maintain Support Vehicle/Equipment Inventory (ICS 218).
- Maintain equipment rental records.
- Maintain equipment service and use records.
- Ensure all equipment time reports are accurate and turned in daily to the Equipment Time Recorder.

Security Manager (SECM)

The Security Manager is responsible for providing safeguards needed to protect personnel and facilities from loss or damage.

Critical Safety Responsibilities

- Establish contacts with local law enforcement agencies. Contact the Liaison Officer or Agency Representatives to discuss any special custodial requirements, which may affect operations.
- Ensure personnel are qualified to manage security problems.
- Develop Security Plan for incident facilities.
- Coordinate security activities with appropriate personnel.
- Provide assistance in personnel problems or emergency situations through coordination with Agency Representatives.

Logistics Guidelines

General

- Locate sleeping areas out of danger from vehicles, aircraft, and other equipment.
- Participate in the developing the Demobilization Plan.
- Control dust.
- Give high priority to environmental protection when locating incident facilities.
- Coordinate locations with the Agency Administrator.
- Develop and post an evacuation plan.
- Inspect facilities for safety and fire hazards on a regular basis and take corrective action where needed.
- Consider need for computer support for resource ordering and inventory, and manage, if provided.

Food Service

Compliance with health and sanitation requirements (OSHA, state, and local) is required in all situations.

- Proper supervision is important to meet food service sanitation requirements.

Water Supply

Select a known, safe water supply or haul in water. Usually it is best to haul in water from a domestic water supply. Otherwise, ensure that it is:

- Adequate, tested, and safe.
- Protected from contamination.

Sanitation Guide

- Local environmental regulations must be met.
- Suggested standards are one standard size (32-gallon) garbage can for every <u>20 persons in an eating area</u> and one can for every 40 persons in other areas.
- Provide adequate toilet facilities, and establish a regular inspection and maintenance schedule to keep them clean.
- Locate toilets properly and treat them to eliminate flies and other insects.
- Suggested standards are one toilet for every 15 to 20 persons, with daily or more frequently scheduled maintenance.
- Transportation
- Post direction signs on roads to facilities and drop points.

Chapter 3 – Position Responsibilities

- Post signs at drop points.
- Carefully plan for transportation of personnel and tools to and from the fireline.
- Provide adequate rest for drivers.
- Isolate fuel storage areas and post signs to them.
- Develop a vehicle control plan, and strictly enforce it.

Communications

Preparing a Communications Plan is the first step towards providing a workable communications system.

Set up incident communications in the following order of priority to meet safety and tactical resource management needs:

1. Communications on fireline: tactical and command nets.
2. Communications between fireline and incident Base.
3. Air operations: ground to air and air to air.
4. Communications between Incident Communications Center and the nearest available service center.
5. Communications in Base/Camp: logistics net.
6. Specialty systems, such as radio telephone interconnect (RTI) (voice), satellite (voice and data), landline telephone (voice and data), Automatic Data Processing (ADP) capability, data transmission by radio.

Communications Plan

A Communications Plan should be prepared for each operational period and should include:

- Incident Radio Communications Plan (ICS 205)
- Telephone facilities
- Number of lines
- Location of telephone(s)

Operating a Fire Communications System

- Use competent, qualified Incident Dispatchers.
- Use clear text in all radio communications.

Coordinating Radio Frequencies

It is very important to maintain system isolation and integrity of the communications system within the incident. Coordination at the regional and national levels is often important to maintain flexibility of all systems within national incident radio support caches. Radio frequencies are a limited resource, and only those required to provide the incident with effective communications should be used.

Procurement

- Coordinate with Procurement Unit Leader in the Finance/Administration Section.

Factors to Consider When Locating and Laying Out an Incident Base or Camp

The Logistics Section Chief should ensure that the following factors are included when assessing potential sites and their subsequent selection.

- Environmental constraints: temporary and permanent effects.
- Ownership of land; written agreement to use site.
- Accessible from existing roads with right of way.
- Communications services available.
- Safety and sanitation, including areas free from smoke.
- Adequate space for facilities, equipment, and people.
- Proximity to fire: safety and travel time.
- Public interference: proximity to and access by the public.
- Water supply: how much, how far, etc.
- Existing facilities: usable, cost, protection needed, etc.
- Potential or planned use of additional Camps.

FINANCE/ADMINISTRATION

Organization Chart

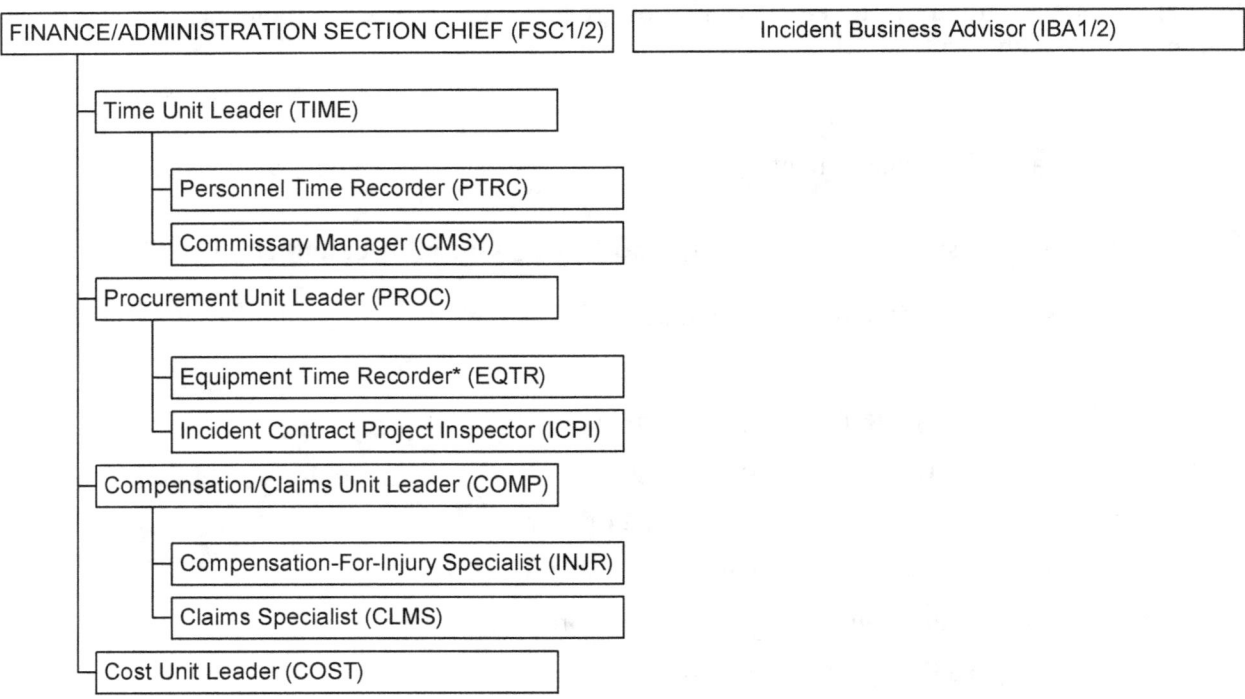

*On some incidents, the Equipment Time Recorder is assigned to and reports to the Procurement Unit Leader; however, this is a skilled position and can be assigned anywhere in the Incident Command organization. Some managers prefer to keep all timekeeping under the Time Unit and assign the Equipment Time Recorder to the Time Unit Leader.

Position Checklists

Finance/Administration Section Chief (FSC1/2)

The Finance/Administration Section Chief is responsible for all financial, administrative, and cost analysis aspects of the incident and for supervising members of the Finance/Administration Section.

- Develop an operating plan for the Finance/Administration Section; fill supply and support needs.
- Review contacts, memoranda of understanding, and cooperative agreements for incident impact and application.
- Determine need for commissary operation.
- Meet with assisting and cooperating agency representatives as required.
- Provide input on financial and cost-analysis matters.

Maintain daily contact with agency(s) administrative headquarters on financial matters.

- Ensure that personnel time records are transmitted to home agencies according to policy.
- Ensure that obligation documents initiated at the incident are properly prepared and completed.
- Before leaving the incident, brief agency administrative personnel on incident-related business management issues needing attention and followup.
- Act as liaison between the IMT and the Incident Business Advisor.

Time Unit Leader (TIME)

The Time Unit Leader is responsible for recording personnel time and managing the commissary operation.

- Determine requirements for the time-recording function.
- Ensure that personnel time-recording documents are prepared daily and comply with agency(s) policy.
- Establish commissary operation as required.
- Establish and maintain adequate records security.
- Before demobilization, release time reports from assisting agency personnel to the respective Agency Representatives.
- Brief Finance/Administration Section Chief on current problems and recommendations, outstanding issues, and followup requirements.
- Determine the need for Personnel Time Recorders, and order personnel as needed (with Finance/Administrative Chief's approval).

- Before demobilization, ensure that all personnel time logs and forms are complete according to agency policy.
- Obtain Demobilization Plan, and ensure that Personnel Time Recorders are adequately briefed on Demobilization Plan.

Personnel Time Recorder (PTRC)

Under supervision of the Time Unit Leader, the Personnel Time Recorder is responsible for overseeing the recording of time for all personnel assigned to an incident.

- Establish and maintain a file for employee time reports within the first operational period.
- Initiate, gather, or update a time report for all personnel assigned to the incident for each operational period.
- Ensure that all employee identification information is verified on the time report.
- Post personnel travel and work hours, transfers, promotions, specific pay provisions, and terminations, to personnel time documents.
- Post all commissary issues to personnel time documents.
- Ensure that time reports are signed.
- Close out time documents before personnel leave the incident.
- Distribute all time documents according to agency policy.
- Maintain a daily log of excessive hours worked and give to Time Unit Leader.

Commissary Manager (CMSY)

Under the supervision of the Time Unit Leader, the Commissary Manager is responsible for commissary operations and security.

- Set up and provide commissary operation to meet incident needs.
- Establish and maintain adequate commissary security.
- Request commissary stock through Supply Unit Leader (must have Finance/Administration Section Chief approval).
- Maintain complete record of commissary stock, including invoices for material received, issuance records, transfer records, and closing inventories.
- Maintain commissary issue records. Submit records to time recorder during or at the end of each operational period.
- Use proper agency forms for record keeping. Complete forms according to agency specification.
- Before demobilization, ensure that all records are closed out and commissary stock is inventoried and returned to Supply Unit.

Procurement Unit Leader (PROC)

The Procurement Unit Leader is responsible for administering all financial matters pertaining to vendor contracts, leases, and fiscal agreements.

- Review incident needs and any special procedures with Unit Leaders, as needed.
- Coordinate with local jurisdiction on plans and supply sources.
- Develop incident procurement procedures for local purchase.
- Prepare and sign contracts and agreements as needed.
- Establish contracts and agreements with local supply vendors as required.
- Ensure that a system is in place that meets agency property management requirements and accounting for all new property purchases.
- Interpret contracts and agreements, and resolve claims or disputes within delegated authority.
- Provide for coordination between the Ordering Manager, agency dispatch, and all other procurement organizations supporting the incident.
- Coordinate with Compensation/Claims Unit on procedures for handling claims.
- Complete final processing of contracts and agreements, and process documents for payment.
- Coordinate cost data, in contracts, with Cost Unit Leader.
- Brief Finance/Administration Section Chief on current problems and recommendations, outstanding issues, and followup requirements.
- Determine the need for Equipment Time Recorders, and order personnel as needed (with Finance/Administration Section Chief's approval).
- Before demobilization, ensure that all procurement logs and forms are completed according to agency policy.
- Obtain the Demobilization Plan, and ensure that the Equipment Time Recorders are adequately briefed on the Demobilization Plan.

Equipment Time Recorder (EQTR)

Under supervision of the Procurement Unit or Time Unit Leader, the Equipment Time Recorder is responsible for overseeing the recording of time for all equipment assigned to an incident.

- Assist Resources, Ground Support, and Facilities Units in establishing a system for collecting equipment time reports.
- Post equipment time after each operational period.
- Prepare a payment document for equipment as required.
- Submit data to supervisor for cost effectiveness analysis as required.
- Maintain current posting on all charges or credits for fuel, parts, services, and commissary.
- Verify all time data and deductions with owner or operator of equipment.
- Complete all forms according to agency specifications.
- Close out forms before demobilization; distribute copies per agency and incident policy.

Incident Contract Project Inspector (ICPI)

The Incident Contract Project Inspector represents the Contracting Officer Representative (COR) and the Contracting Officer's Technical Representative (COTR) to secure compliance with terms and conditions of the contract and to notify the contractor of any deviations from contract requirements.

- Provide subject matter expertise in the field to the incident supervisor of contracted equipment (e.g., Equipment Manager, Equipment Inspector, Operations personnel) to ensure contract requirements are met.
- Consult with the COR or COTR concerning matters that may require contract action.

Compensation/Claims Unit Leader (COMP)

The Compensation/Claims Unit Leader is responsible for the overall management and direction of all administrative matters pertaining to compensation-for-injury and claims-related activities related to an incident.

- Establish contact with Safety Officer, Liaison Officer, and Agency Representatives.
- Coordinate with Interagency Resource Representative, if any are assigned.
- Establish a compensation-for-injury work area within or as close as possible to the Medical Unit.
- Determine the need for Compensation-For-Injury and Claims Specialists, and order personnel as needed.
- Review Incident Medical Plan.
- Coordinate with Procurement Unit on procedures for handling claims.

- Periodically review logs and forms produced by the Compensation-For-Injury and Claims Specialists to ensure compliance with agency requirements and policies.
- Obtain Demobilization Plan, and ensure that the Compensation-For-Injury and Claims Specialists are adequately briefed on the Demobilization Plan.
- Before demobilization, ensure that all compensation-for-injury and claims logs and forms are complete and routed to the appropriate agency for postincident processing.

Compensation-For-Injury Specialist (INJR)

Under the supervision of the Compensation/Claims Unit Leader, the Compensation-For-Injury Specialist is responsible for administering financial matters resulting from serious injuries and fatalities occurring on an incident. Close coordination is required with the Medical Unit.

- Colocate compensation-for-injury operations with those of the Medical Unit when possible.
- Establish procedure with Medical Unit Leader for prompt notification of injuries or fatalities.
- Establish contact with Safety Officer and Agency Representatives.
- Obtain copy of Medical Plan (ICS 206).
- Provide written authority, according to agency policy, for persons requiring medical treatment.
- Ensure that correct agency forms are used.
- Provide correct billing forms for transmittal to doctor and hospital.
- Keep informed, and report on status of hospitalized personnel.
- Obtain all witness statements from Safety Officer and Medical Unit, and review for completeness.
- Coordinate the analysis of injuries with the Safety Officer.
- Maintain log of all injuries occurring on the incident.
- Coordinate with appropriate agency(s) to look after injured personnel in local hospitals after demobilization.

Claims Specialist (CLMS)

Under the supervision of the Compensation/Claims Unit Leader, the Claims Specialist is responsible for managing all claims-related activities (other than injury) for an incident.

- Develop and maintain a log of potential claims.
- Initiate claim investigations.
- Request skilled investigation from appropriate agency, when needed.
- Ensure site and property in investigation are protected.
- Coordinate with investigation team as necessary.
- Obtain witness statements pertaining to claims.
- Review investigations for completeness and followup action needed by local agency.
- Keep the Compensation/Claims Unit Leader advised on existing and potential claims.
- Ensure use of correct agency forms.
- Document any incomplete investigations.

Cost Unit Leader (COST)

The Cost Unit Leader is responsible for collecting all cost data, performing cost-effectiveness analyses, and providing cost estimates and cost-saving recommendations.

- Coordinate with agency on cost-reporting procedures.
- Collect and record all cost data.
- Prepare incident cost summaries.
- Prepare resource-use cost estimates for Planning Section.
- Recommend cost savings to Finance/Administration Section Chief.
- Maintain cumulative incident cost records.
- Complete all records before demobilization.
- Provide reports to Finance/Administration Section Chief.

CHAPTER 4 – REFERENCE

PORTABLE PUMPS AND HYDRAULICS

When considering the use of portable pumps and hose lays during fire suppression activities, it is important to size up the situation and do some hydraulics calculations to determine where and when to use a portable pump. Some items to consider are pump capability needed, adequacy of water source, and the type of hose lay to use.

In determining required pumping capacity, it is necessary to consider factors, such as friction loss due to length and diameter of hose, desired nozzle pressure, number of nozzles, type of nozzle, and head pressure.

Formula for Determining Pump Pressure

Note: All references to pressure (pump pressure, nozzle pressure, head gain or loss, friction loss, etc.) is in pounds per square inch (psi).

PDP = NP + or - H + FL where:

PDP = Pump Discharge Pressure at the discharge side of the pump.

NP = Pressure required at the nozzle for the most efficient operation.

Remember: The larger the nozzle tip, the more PDP (pump pressure) is needed to maintain a given nozzle pressure.

H = Head. Add (+) if pumping uphill and subtract (-) if pumping downhill.

Remember: One (1) psi will raise (lift) water approximately 2 feet in elevation. Conversely, for every 2-foot drop in elevation, about 1 psi will be gained. Head pressure is only determined by change in elevation and is independent of hose size.

FL = Friction Loss

Remember: The smaller the hose, the higher the friction loss; the larger the hose, the lower the friction loss. For example, a 1-inch hose has about seven times the friction loss as a 1½-inch hose.

Reminders for Using Portable Pumps and Hose Lays

- A pump can be ruined in minutes if proper operational procedures are not followed.
- Do not use a longer suction hose than necessary.
- For good priming, remove any humps in the suction hose.
- Be alert for cavitation in the pump if you notice the sound of rolling marbles or see an increase in rpm without a corresponding increase in pressure. This is often a result of insufficient water supply to meet the demand.
- Keep your pump as close to your water source elevation as possible. The maximum practical vertical lift (water source to the pump) for most pumps is 22 feet at sea level. As lift is increased, pump performance decreases.
- Protect your pump from ingesting sand, silt, or gravel by using a suction strainer or foot valve and putting the suction hose intake in a pail or on a shovel.
- Minimum working nozzle pressure is about 25 psi, but the recommended minimum is 50 psi.
- Use a check and bleeder valve, a shutoff valve, or gated wye valve near the pump on the <u>discharge</u> side when pumping uphill to prevent draining your hose lay (by backflow) when the pump is not running. **Note**: A check-and-bleeder valve operates automatically, whereas a shutoff valve or gated wye must be operated manually.
- Know your water source. Some natural sites may have other water-use demands. Water levels and flows can change on a daily basis. Check the pump site and available water levels before conducting continuous operations.

Drafting Guidelines

Maximum practical lift with a good serviceable pump = 22 ft of lift at sea level
= 14 ft of lift at 8,000 ft elevation

Expected Output of Commonly Used Portable Pumps at Sea Level

Pump Type	Pressure (psi)	Flow (gal/min)
Waterous Floto-Pump	150	20
High Pressure Pump (Mark 3, Wick 375)	150	65
Honda WX10	Free flow	37
Mini Striker	50	32

General Rules for Fireline Hydraulics

Pressure	
1 psi	= 2 ft of water lift (2.31 ft actual)
2 ft of head	= 1 psi (0.86 psi actual)
50 ft of head	= 25 psi (22 ft actual)
100 ft of head	= 50 psi (43 ft actual)
Atmospheric pressure	= 14.7 psi at sea level (29.92 in of mercury (Hg))
1,000-ft elevation gain	= ½ psi decrease in atmospheric pressure = 1 in of mercury (Hg) decrease in atmospheric pressure (1-in loss of vacuum) = Decrease in drafting of 1 ft of water lift
Drafting	
1 in of mercury (Hg)	= 1 ft of water lift (1.13 ft actual) = 22 ft of lift at sea level = 15 ft of lift at 8,000 ft
Nozzle Pressure	
Tips (except master streams)	= 50 psi
Region 5 Forester (twin tip) nozzle	= 50 psi
Combination (fog) nozzle	= 100 psi
Water Weight and Volumes	
Weight of water	= 8 lb/gal (8.33 lb/gal actual)
Volume of water in 100-ft length of fire hose	= 4 gal for 1-in hose = 9 gal for 1½-in hose

Chapter 4 – Reference

Friction Loss by Hose Size and Type

Friction Loss per 100 Feet for Straight-Stream Tips and Combination (Fog) Nozzles

Straight Stream Tips (Nozzle Pressure = 50 psi)						
Tip Orifice Size (in)	1/8	3/16	1/4	5/16	3/8	1/2
Flow (gal/min)	3	7	13	21	30	53
Hose Diameter (in)	**Friction Loss (psi)**					
5/8	2	11	34	84	180	551
3/4	1	6	19	46	99	303
1	0	1	4	11	23	69
1½	0	0	1	1	3	10

Friction Loss Per 100 ft (psi) applies to the left-side label.

Combination (Fog) Nozzles (Nozzle Pressure = 100 psi)						
Flow (gal/min)	10	20	30	60	90	100
Hose Diameter (in)	**Friction Loss (psi)**					
5/8	20	80	180	720	1,620	2,000
3/4	11	44	99	396	891	1,100
1	3	10	23	90	203	250
1½	0	1	3	13	28	35

Pump Pressures for 50-psi Nozzle Pressure

1-inch Hose

Tip Orifice Size (in)		1/8	3/16	1/4	5/16	3/8	1/2
Flow (gal/min)		3.3	7.41	13.2	20.7	29.7	53
Friction Loss Per 100 ft (psi)		0	1.4	4.4	10.7	22.1	70.2
Hose Length (ft)	Nozzle Above Pump (ft)	\multicolumn{6}{c}{Required Pump Pressure (psi)}					
100	0	50	51	54	61	72	120
100	100	94	95	97	104	115	164
200	0	51	53	59	71	94	190
200	100	94	96	102	115	137	234
300	0	51	54	63	82	116	261
300	100	94	97	106	125	159	304
300	200	137	141	150	169	203	347
400	0	51	55	67	93	138	331
400	100	94	99	111	136	181	374
400	200	138	142	154	179	225	417
400	300	181	185	197	223	268	461
500	0	51	57	72	104	160	401
500	100	95	100	115	147	204	444
500	200	138	143	158	190	247	488
500	300	181	187	202	233	290	531
1,000	0	53	64	94	157	271	752
1,000	100	96	107	137	200	314	796
1,000	200	139	150	163	244	357	839
1,000	300	183	194	223	287	400	882
1,000	400	226	237	267	330	444	925
1,000	500	269	280	310	374	487	969
1,000	600	312	323	353	417	530	1,012

Chapter 4 – Reference

Pump Pressures for 50-psi Nozzle Pressure

1½-Inch Hose

Tip Orifice Size (in)		1/8	3/16	1/4	5/16	3/8	1/2
Flow (gal/min)		3.3	7.41	13.2	20.7	29.7	53
Friction Loss Per 100 ft (psi)		0	0.2	0.6	1.5	3.1	9.8
Hose Length (ft)	Nozzle Above Pump (ft)	\multicolumn{6}{c}{Required Pump Pressure (psi)}					
100	0	50	50	51	51	53	60
100	100	93	93	94	95	96	103
200	0	50	50	51	53	56	70
200	100	93	94	95	96	99	113
300	0	50	51	52	54	59	79
300	100	93	94	95	98	103	123
300	200	137	137	138	141	146	166
400	0	50	51	52	56	62	89
400	100	93	94	96	99	106	133
400	200	137	137	139	143	149	176
400	300	180	181	182	186	192	219
500	0	50	51	53	57	65	99
500	100	93	94	96	101	109	142
500	200	137	138	140	144	152	186
500	300	180	181	183	187	195	229
1,000	0	50	52	56	65	81	148
1,000	100	94	95	99	108	124	192
1,000	200	137	139	143	152	167	235
1,000	300	180	182	186	195	211	278
1,000	400	224	225	229	238	254	321
1,000	500	267	268	273	281	297	365
1,000	600	310	312	316	325	341	408

Pump Pressures for 50-psi Nozzle Pressure (Continued)

1½-Inch Hose

Tip Orifice Size (in)	1/8	3/16	1/4	5/16	3/8	1/2
Flow (gal/min)	3.3	7.41	13.2	20.7	29.7	53
Friction Loss Per 100 ft (psi)	0	0.2	0.6	1.5	3.1	9.8
Hose Length (ft)	Nozzle Above Pump (ft)	Required Pump Pressure (psi)				

Hose Length (ft)	Nozzle Above Pump (ft)					
2,000	0	51	54	62	80	112	247
2,000	100	94	97	105	123	155	290
2,000	200	137	140	149	167	198	333
2,000	300	181	184	192	210	242	377
2,000	400	224	227	235	253	285	420
2,000	500	267	270	279	296	328	463
2,000	600	311	314	322	340	371	506
2,000	700	354	357	365	383	415	550
2,000	800	397	400	409	426	458	593
3,000	0	51	56	68	95	143	345
3,000	100	94	99	112	138	186	388
3,000	200	138	142	155	182	229	432
3,000	300	181	186	198	225	272	475
3,000	400	224	229	241	268	316	518
3,000	500	268	272	285	311	359	561
3,000	600	311	316	328	348	402	605
3,000	700	354	359	371	398	446	648
3,000	800	397	402	415	441	489	691

Chapter 4 – Reference

FOAM

Foam Use

Low-expansion foams have proven to be valuable in the suppression of fire by increasing the effectiveness of water.

- Foam solution can be used effectively with regular nozzles but is most effective with air-aspirating nozzles or a compressed air foam system (CAFS).
- Foam has the ability to adhere to and cool fuels for a much longer period of time than water. In addition, foam allows time for water to be absorbed by the fuels.
- Rates of application (including width and depth) depend upon wind, temperature, fuel moisture, and fuel loading.
- In general, enough foam is required to fully coat exposed fuels and to sufficiently raise fuel moistures.

Foam Mixture Rates

A 0.3 mixture (0.3 gallons of foam concentrate to 100 gallons of water) is the average recommended for most situations regardless of the system being used (compressed air, air-aspirating nozzles, or regular nozzles). However, mixture rates may vary from 0.1 of 1% used during mop up to a full 1% for structure protection.

Note: More concentrate may be required if the water has a high mineral content, but the mixture rate should never exceed 1%.

Mixture Rated By Application and Type of Equipment			
	Foam-to-Water Mixture, in %		
Application	Compressed Air System	Air-Aspirating Nozzle	Regular Nozzle
Direct Attack	0.3	0.3–0.5	0.3–0.5
Indirect Attack	0.3	0.3–0.5	0.3–0.5
Mop up	0.3	0.3–0.5	0.3–0.5
Structures	0.3	0.3–0.5	0.5

Foam for Direct Attack

- Place foam directly at the base of the flame.
- Use foam to coat burning materials. Leave a foam blanket over hot fuels to continue wetting the fuels.
- When attacking the fire's edge, also apply foam onto adjacent unburned fuels.

Foam for Indirect Attack

- Apply the foam directly in advance (within 5 feet) of the person setting the backfire.
- The foam line should be at least 2½ times as wide as the average flame height.
- Coat all sides of fuel when possible.
- The foam line can be reinforced and widened on the up wind side once the original control line has been established and backfiring or burnout has begun.

Foam for Mop Up

- For best penetration, apply foam solution as you would a water stream.
- Use a high-pressure wet water mist to create a frothy foam for close in mop up. This works extremely well on pitchy or punky material, duff, and litter.
- A mop up wand is very effective with foam solution for deep-seated fires in stumps, landings, log decks, etc.
- "Forester" (See Region 5 Forester under Nozzle Pressure.) nozzles also work well with foam solution in mop up.

Foam for Exposure Protection

- Foam is most effective when applied shortly before heat exposure. Apply enough foam in advance of the fire to allow penetration, yet not so long that the foam evaporates and dissipates. In general, foam applied by a compressed air system will last about 1 hour, and foam applied by an air-aspirated nozzle will last about 30 minutes in hot weather.
- High-quality foaming agents will leave at least ½ inch of foam on all surfaces.
- Make the foam line 2 times as wide as the flame length when creating a foam line for backfiring or burning out.
- When coating unburned fuels, use a wet foam that will penetrate and soak fuels down to the soil.
- Foam is most effective when applied immediately before ignition.
- Coat exposed vertical fuels as high as the system being used will reach.

- Use foam that clings to a vertical surface when protecting trees, snags, log decks, telephone poles, etc. Sufficient time must be allowed to thoroughly coat these fuels. Apply foam in a radius 2 times the height of standing objects to be protected.

- Apply foam to the outside walls, eaves, roofs, columns, or other threatened surfaces when protecting structures. Loft foam from a distance far enough away to avoid foam breakdown.

Foam Safety

- Maintain communications between the nozzle operator and the engine using radio or hand signals.
- Avoid contact with skin and clothes.
- Gloves and eye protection should be worn.
- If foam or foam solution gets into eyes, irrigate with water immediately.
- Follow the safety guidelines on the foam container.
- The use of CAFS requires special training.
- Use caution, as any surface covered with foam can be very slippery.

USE OF FIRELINE EXPLOSIVES

Advantages

- Rapid line construction with minimal personnel needs.
- Work well in steep, difficult terrain where fuels are light to moderate.
- Brush and debris is scattered rather than piled next to the line.
- Soil is loosened to facilitate line improvement and hotspotting.
- Line width is easily varied by the number of strands of explosive used.
- Produce a more environmentally acceptable fireline.

Disadvantages

- Limited availability of trained and experienced personnel.
- Requires that all personnel working on the fire be accounted for and removed from the blasting area.
- Transporting the explosives presents unique problems.
- The need to provide security.
- Fireline explosives are becoming more expensive.

Note: Refer to the Fireline Explosives Production Comparisons table. Productivity comparison charts for explosives appear later in this chapter.

HAZMAT CHECKLIST FOR INCIDENT BASE MANAGEMENT

- Be able to identify what materials may be classified as hazardous.
- Be familiar with transportation and storage of HazMat.
- Make sure HazMat storage areas have been selected and posted clearly in Camps or Bases.
- Know local HazMat contacts and waste disposal sites, etc.
- Inform the Supply Unit Leader that this position has the responsibility for HazMat while in a Camp setting as well as for HazMat being demobilized.
- It is critical that Supply Unit Leaders are in communication with cache personnel <u>when ordering</u> and <u>returning</u> hazardous materials. Cache Demobilization Specialists can be resource ordered or contacted for the proper handling and returning of any HazMat.
- The Demobilization Plan needs to include specific instructions by the Supply Unit Leader for returning all hazardous materials to:
    - ✓ Cache(s)
    - ✓ Local host agency(s)
    - ✓ Local HazMat contractors
    - ✓ Hazardous waste disposal site

USE OF INMATE CREWS

Some states have access to inmate labor for fire operations. Situations may arise where inmates are used on fires involving personnel from many agencies.

Although each state has specific rules governing the use of inmates, the following guidelines will apply in most situations. Check with the inmate crew Liaison Officer, the officer-in-charge, or the appropriate AREP for more specific information in your area.

- Crews on fireline are supervised by forest crew supervisors (resource boss or higher).
- Inmate crews are usually limited to use within the state where they are based although some states have interstate agreements with neighboring states.
- Contact with inmates should be done through the corrections officer-in-charge in camp.
- Contact with inmates should be done through the forest crew supervisor on the fireline.
- Consult the officer-in-charge before giving direction to inmates.
- Keep relationships with inmates on a business basis. For example, do not play cards with, carry messages for, bring gifts to, accept gifts from, or make purchases for the inmates.
- The officer-in-charge or other inmate Camp representative may act as liaison with Fire Overhead on all matters pertaining to inmates (food, bedding areas, etc.).
- The officer-in-charge will remain with the crew while on the fireline. Any fire suppression related problems, such as pumps, tools, drinking water, fire equipment, etc., are to be taken care of by Fire Overhead.
- Inmates should not be used in a "Squad Boss" type position, or given supervision over fellow inmates.
- Inmate crews should be provided a separate sleeping area where they can be away from other crews.
- Provide separate sleeping areas for male and female, and adult and juvenile, crews.
- Interspersing inmate crews with civilian crews on the fireline is generally permitted (but not encouraged), provided the crew supervisor is aware of the situation at all times.
- Intermingling of inmates at the incident Base with civilians should only occur at meal times.
- Inmates will be confined to the incident Base or Camp while offshift.
- Inmates shall not be allowed to handle explosives or detonating devices.
- Civilians and inmates shall have separate schedules for bathing.

PRODUCTION TABLES

Sustained Line Production Rates of 20-Person Crews in Feet per Hour*

Fire Behavior Fuel Model	Type I Direct	Type I Indirect	Type II & II IA Direct	Type II & II IA Indirect
1 Short Grass 2 Open Timber Grass	**1,122** (792–1,386)**	**627** (508–746)	**627** (174–660)	**285** (174–380)
4 Chaparral	**436** (330–528)	**330** (178–482)	**449** (80–640)	**272** (178–376)
5 Brush	**1,089** (924–1,254)	**323** (244–403)	**471** (304–682)	**277** (178–376)
6 Dormant Brush Hardwood Slash	**1,089** (924–1,254)	**323** (244–403)	**471** (304–682)	**277** (178–376)
8 Closed Timber Litter 9 Hardwood Litter 10 Timber Litter & Understory	**693** (594–792)	**455** (396–515)	**447** (370–448)	**378** (255–452)

*Based on San Dimas Technology & Development Center, Tech Tip – 1151-1805P. Fireline Production Rates, 2011. No data was collected in fuel models 3, 7, and 11 – 13.

**Numbers in parentheses are expected ranges of line production.

IA = Initial Attack

Sustained Line Production Rates of 20-Person Crews in Feet per Hour*

Fire Behavior Fuel Model	Crew Type 1	Crew Type 2
7 Southern Rough	264	132
11 Logging Slash, Light	990	594
12 Logging Slash, Medium	462	264
13 Logging Slash, Heavy	330	198

*Based on various sources from pre-1980.

Chapter 4 – Reference

Sustained Line Production Rates of 20-Person Crews in Chains per Hour*

Fire Behavior Fuel Model	Type I Direct	Type I Indirect	Type II & II IA Direct	Type II & II IA Indirect
1 Short Grass 2 Open Timber Grass 3 Tall Grass	17 (12–21)**	9.5 (7.7–11.3)	10.0 (5.0–15.0)	4.2 (2.7–5.7)
4 Chaparral	6.6 (5–8)	5 (2.7–7.3)	7.0 (6.2–7.9)	4.2 (2.7–5.7)
5 Brush	16.5 (14–19)	4.9 (3.7–6.1)	7.0 (6.2–7.9)	4.2 (2.7–5.7)
6 Dormant Brush Hardwood Slash	16.5 (14–19)	4.9 (3.7–6.1)	7.0 (6.2–7.9)	4.2 (2.7–5.7)
8 Closed Timber Litter 9 Hardwood Litter 10 Timber Litter & Understory	10.5 (9–12)	6.9 (6.0–7.8)	7.0 (6.2–7.9)	4.2 (2.7–5.7)

*Based on San Dimas Technology & Development Center, Tech Tip – 1151-1805P, Fireline Production Rates, 2011.

**Numbers in parentheses are expected ranges of line production.

IA = Initial Attack

Sustained Line Production Rates of 20-Person Crews in Chains per Hour*

Fire Behavior Fuel Model	Crew Type I	Crew Type II
7 Southern Rough	4	2
11 Logging Slash, Light	15	9
12 Logging Slash, Medium	7	4
13 Logging Slash, Heavy	5	3

*Based on various sources from pre-1980.

Line Production Rates for Initial Action by Hand Crews in Chains per Person per Hour

Fire Behavior Fuel Model	Specific Conditions	Construction Rate (in chains per person per hour)
1 Short Grass	Grass Tundra	4.0 1.0
2 Open Timber/Grass Understory	All	3.0
3 Tall Grass	All	0.7
4 Chaparral	Chaparral High Pocosin	0.4 0.7
5 Brush	All	0.7
6 Dormant Brush/Hardwood Slash	Black Spruce Others	0.7 1.0
7 Southern Rough	All	0.7
8 Closed Timber Litter	Conifers Hardwoods	2.0 10.0
9 Hardwood Litter	Conifers Hardwoods	2.0 8.0
10 Timber (Litter & Understory)	All	1.0
11 Logging Slash, Light	All	1.0
12 Logging Slash, Medium	All	1.0
13 Logging Slash, Heavy	All	0.4

Note: These rates are to be used for estimating initial action productivity only. Do not use these rates to estimate sustained line construction, burnout, and holding productivity. Initial action may consist of scratch line construction and hotspotting.

Chapter 4 – Reference

Line Production Rates for Initial Action by Engine Crews in Chains per Crew per Hour

| Fire Behavior Fuel Model | Specific Conditions | Chains Per Crew Hour |||||
| | | Number of Persons in Crew |||||
		1	2	3	4	5+
1 Short Grass	Grass	6	12	24	35	40
	Tundra	2	8	15	24	30
2 Open Timber/Grass Understory	All	3	7	15	21	25
3 Tall Grass	All	2	5	10	14	16
4 Chaparrel	Chaparrel	2	3	8	15	20
	High Pocosin	2	4	10	15	18
5 Brush (minimum 2 ft tall)	All	3	6	12	16	20
6 Dormant Brush/Hardwood Slash	Black Spruce	3	6	10	16	20
	Others	3	6	12	16	20
7 Southern Rough	All	2	5	12	16	20
8 Closed Timber Litter	Conifers	3	8	15	20	24
	Hardwoods	10	30	40	50	60
9 Hardwood Litter	Conifers	3	7	12	18	22
	Hardwoods	8	25	40	50	60
10 Timber (Litter & Understory)	All	3	8	12	16	20
11 Logging Slash, Light	All	3	8	12	16	20
12 Logging Slash, Medium	All	3	5	10	16	20
13 Logging Slash, Heavy	All	2	4	8	15	20

Note: These rates are to be used for estimating initial action productivity only. <u>Do not</u> use these rates to estimate sustained line construction, burnout, and holding productivity. Initial action may consist of scratch line construction and hotspotting.

Fireline Explosives Production Comparisons

Production Rate Comparison between a 7-Person Fireline Explosives Crew and a 20-Person Hand Crew over a 10-Hour Shift

Fuel Type	Constructed Fireline (in chains)	
	Explosives Crew	Hand Crew
Grass	360	360
Second-Growth Conifers	240	180
Light Slash	210	90
Heavy Slash	120	45

Note: This is based upon Washington State Department of Natural Resources experience.

Dozer Fireline Construction Rates (Single Pass) in Chains per Hour

Fire Behavior Fuel Model	Up or Down Slope	Slope Class 1 0–25%	Slope Class 2 26–40%	Slope Class 3 41–55%	Slope Class 4 56–74%
Type III Dozer 1, 2	Up Down	55–90 90–110	30–55 90–110	8–30 20–90	0–8 0–20
3, 5, 8	Up Down	45–70 70–80	25–45 65–80	2–25 0–65	0–2 0
4.00	Up Down	20–35 35–40	10–20 25–40	0–10 0–25	0 0
6, 7, 9	Up Down	35–55 55–60	15–35 40–60	0–15 0–40	0 0
11, 12	Up Down	15–25 25–30	7–15 10–30	0–7 0–10	0 0
10, 13	Up Down	8–15 10–15	3–8 5–10	0–3 0–5	0 0
Type II Dozer 1, 2	Up Down	85–125 125–145	60–85 130–145	30–60 75–130	0–30 0–75
3, 5, 8	Up Down	70–105 105–120	45–70 105–120	15–45 55–105	0–15 0–55
4.00	Up Down	35–60 60–75	20–35 65–76	2–20 20–65	0–2 0–20
6, 7, 9	Up Down	50–85 85–100	30–50 85–100	7–30 40–85	0–7 0–40
11, 12	Up Down	25–40 40–55	15–25 45–55	1–15 0–45	0–1 0
10, 13	Up Down	10–20 20–25	7–10 20–25	0–7 0–20	0 0
Type I Dozer 1, 2	Up Down	100–140 140–155	70–100 140–155	35–70 85–140	0–35 0–85
3, 5, 8	Up Down	75–110 110–130	50–75 110–130	20–50 55–110	0–20 0–55
4.00	Up Down	45–70 70–80	30–45 75–85	8–30 25–75	0–8 0–25
6, 7, 9	Up Down	65–95 95–110	40–65 90–110	15–40 50–90	0–15 0–50
11, 12	Up Down	35–55 55–65	20–35 55–65	3–20 6–55	0–3 0–6
10, 13	Up Down	20–35 35–40	9–20 30–40	0–9 0–30	0 0

Dozer Fireline Construction Rates (Single Pass) in Chains Per Hour (Continued)

Note: Production rates are not precise but vary with conditions. The higher rate can be applied for situations involving:

- Newer dozers (1975 and later)
- Dozers in excellent operating condition
- Most-qualified operators
- Temperatures below 90 °F
- Moist soil, few or no rocks
- No lost time
- Indirect fireline
- Average fire behavior
- Daylight operations
- Less resistive vegetative types within each fire behavior fuel model

Dozer	Horse Power	Examples
Type I	HEAVY 200 Minimum Horse Power	D-8, D-7, JD-950
Type II	MEDIUM 100 Minimum Horse Power	D-5N, D-6N, JD-750
Type III	LIGHT 50 Minimum Horse Power	JD-450, JD-550, D-3, D-4

Minimum standards for personnel with dozers will differ depending on fuel type, terrain, and resource configuration. Dozer strike teams may use team leader in place of additional personnel per dozer. Fuel requiring burnout and terrain that requires scouting demands two personnel per dozer.

Tractor Plow Fireline Production Rates in Chains per Hour

Drag or Mounted Plow, Appropriate Blade, Level to Rolling Terrain

Fire Behavior Fuel Model	Tractor Plow Type					
	1	2	3	4	5	6
	(165 HP) D-7, JD-850 & Larger	(140 HP) D-6, JD-750, TD-15, Case 1450	(120 HP) D-5H, D-4H, Case 1150	(90 HP) D-4, JD-650, D-5C	(70–80 HP) JD-450, D-4C	(42–60 HP) JD-350, D-3, JD-400
1	240	240	240	200	180	80
2	180	180	180	140	120	80
3	180	180	180	120	100	70
4	80	80	60	40	20	0
5	160	160	160	100	80	40
6	120	120	100	60	40	20
7	160	160	160	120	100	60
8	180	180	180	120	100	70
9	180	180	180	120	100	70
10	100	100	80	50	40	20
	Mountainous terrain, 60% or less slope, front- and rear-mounted plow, downhill plowing					
8	—	—	—	50	40	20
9	—	—	—	50	40	20
	Mountainous terrain, 60% or less slope, using ripper attachment, up/down slope fireline construction					
1, 2, 3	20/30	10/30	0/30	—	—	—
4, 6, 12, 13	10/20	5/10	0/5	—	—	—
5, 7, 8–10, 11	12/25	8/15	0/10	—	—	—

— = Not applicable

INTERAGENCY CREW QUALIFICATIONS AND EQUIPMENT STANDARDS
Minimum Crew Standards for National Mobilization

Minimum Standards	Type 1[1]	Type 2 with Initial Attack Capability	Type 2
Fireline Capability	Initial Attack – Can be broken up into squads, fireline construction, complex firing operations (backfire)	Initial Attack – Can be broken up into squads, fireline construction, firing to include burnout	Initial Attack – Fireline construction, firing as directed
Crew Size	18–20		
Leadership Qualifications	Permanent supervision: Supt.: TFLD, ICT4, FIRB Asst. Supt.: STCR, ICT4 3 Squad Bosses: ICT5 2 Senior Firefighters: FFT1	Crew Boss: CRWB 3 Squad Bosses: ICT5	Crew Boss: CRWB 3 Squad Bosses: FFT1
Language Requirement	All senior leadership, including Squad Bosses and higher, must be able to read and interpret the language of the crew as well as English.		
Experience	80% 1 season	60% 1 season	20% 1 season
Full-Time Organized Crew	Yes (work and train as a unit 40 h per week)	No	No
Communications	5 programmable radios		4 programmable radios
Sawyers	3 agency-qualified	3 agency-qualified	None
Training	As required by the *Interagency Hotshot Crew Guide* or agency policy before assignment	Basic firefighter training and/or Annual Fireline Safety Refresher Training before assignment	Basic firefighter training and/or Annual Fireline Safety Refresher Training before assignment
Logistics	Crew level agency purchasing authority	No purchasing authority	No purchasing authority
Maximum Weight	5,300 lb		
Dispatch Availability	Available nationally	Available nationally	Variable
Production Factor	1.0	.8	.8
Transportation	Own transportation	Transportation needed	Transportation needed
Tools & Equipment	Fully equipped	Not equipped	Not equipped
Personal Gear	Arrives with: Crew first aid kit, personal first aid kit, headlamp, 1-qt canteen, web gear, sleeping bag		
Personal Protective Equipment (PPE)	All standard designated fireline PPE		
Certification	Must be annually certified by the local host Unit Agency Administrator or designee before being made available for assignment.	N/A	N/A

[1] An Interagency Hotshot Crew (IHC) is a Type I crew that exceeds the Type I standards as required by the *Interagency Hotshot Crew Operations Guide* (2011) in the following categories:
- Permanent supervision with 7 career appointments (Superintendent, Assistant Superintendent, 3 Squad Bosses).
- IHCs work and train as a unit 40 hours per week.
- IHCs are a national resource.

N/A = Not applicable

Chapter 4 – Reference

NWCG Engine and Water Tender Typing (Minimum Requirements)

Requirements	Engine Type						
	Structure		Wildland				
	1	2	3	4	5	6	7
Tank minimum capacity (gal)	300	300	500	750	400	150	50
Pump minimum flow (gal/min)	1,000	500	150	50	50	50	10
At rated pressure (psi)	150	150	250	100	100	100	100
Hose: 2½-inch	1,200	1,000	—	—	—	—	—
1½-inch	500	500	1,000	300	300	300	—
1-inch	—	—	500	300	300	300	200
Ladders per NFPA 1901	Yes	Yes	—	—	—	—	—
Master stream 500 gal/min.	Yes	—	—	—	—	—	—
Pump and roll	—	—	Yes	Yes	Yes	Yes	Yes
Maximum GVWR (lb)	—	—	—	—	26,000	19,500	14,000
Personnel (minimum)	4	3	3	2	2	2	2

— = Not applicable
NFPA = National Fire Protection Association
GVWR = gross vehicle weight rating

Requirements	Water Tender Type				
	Support			Tactical	
	S1	S2	S3	T1	T2
Tank capacity (gal)	4,000	2,500	1,000	2,000	1,000
Pump minimum flow (gal/min)	300	200	200	250	250
At rated pressure (psi)	50	50	50	150	150
Maximum refill time (minutes)	30	20	15	—	—
Pump and roll	—	—	—	Yes	Yes
Personnel (minimum)	1	1	1	2	2

— = Not applicable

Note:

1. All types shall meet Federal, state, and agency requirements for motor vehicle safety standards, including all gross vehicle weight ratings (GVWR) when fully loaded.
2. Type 3 engines and tactical water tenders shall be equipped with a foam proportioner system.
3. All water tenders and engine Types 3 through 6 shall be able to prime and pump water from a 10-foot lift.
4. Personnel shall meet the qualification requirements of NWCG's *National Interagency Incident Management System: Wildland Fire Qualification System Guide* (PMS 310-1, June 2012).

Common Additional Needs for Engines and Tenders (Request As Needed)

- All-wheel drive (includes four-wheel drive)
- High-pressure pump (250 psi at one-half flow of Type)
- Foam proportioner
- Compressed Air Foam System (CAFS) 40 ft3/min minimum
- Additional personnel

Air Tankers

Resource	Components	Minimum Standards for Type			
		Type 1	Type 2	Type 3	Type 4
Air Tankers	Minimum capacity (gal)	3,000	1,800–2,999	800–1,799	Less than 800
	Examples:	P-3	DC-6	S-2F	Air Tractor
		DC-7	P2-V	AT-802F	Dromader
		C-130 (MAFFS)*		CL-215/415	Thrush

*Modular Airborne Fire Fighting System

Note: Tanker capacity standards may vary by agency.

Adapted from the *Interagency Aerial Supervision Guide* (PMS 505, January 2011).

Helicopters

Components	Type 1	Type 2	Type 3
Allowable payload at 59 °F at sea level	5,000	2,500	1,200
Passenger seats	15 or more	9–14	4–8
Retardant or water-carrying capability (gal)	700	300	100
Maximum gross takeoff and landing weight (lb)	12,501+	6,000–12,500	Up to 6,000
Examples	Bell 214	Bell 204, 205, 212	Bell 206
Helitanker	• Fixed tank • Certified by Air Tanker Board • 1,100 minimum gallon capacity		

OTHER REFERENCES

Clear-Text Guide

Words and Phrases	Application – Examples
Standard Replies	
Affirmative	Yes
Can handle	Used with the amount of equipment needed to handle the incident. Example: "Waverly 3 can handle with units now at scene."
Copy, Copies	Used to acknowledge message received. Example: "Engine 3 copies."
Disregard	Self-explanatory
Proceed	Indicates another unit may transmit. Example: "Go ahead Essex 50."
How do you copy?	Request for report on transmission quality.
Loud and clear	Self-explanatory
Negative	No
Repeat	Self-explanatory
Standby	Self-explanatory
Unreadable	Signal received is not clear.
Status Reporting	
At scene	Used when units arrive at the scene of an incident.
Assigned	State where and what the assignment is.
Available (location)	Ready to respond to calls. Location is optional.
Available at residence	Used to indicate personnel are available and oncall at home.
Available at scene	No longer needed at scene and are available to respond to other calls.
En route (location)	Used to designate a nonemergency destination. En route is not substitute for responding.
In-quarters (location)	Used to indicate that a resource is at station. Example: "Engine 7 in quarters, Charlottesville."
In-service	Unit is operating, but not in response to a dispatch.
Off duty (location)	Used to sign off when going off duty and are unavailable for calls.
Out-of-contact (location)	Indicates unit is still on duty, but out of radio contact at the location specified.
Out-of-service (location is optional)	Indicates unit is not available due to mechanical problems.
Respond, responding	Used in dispatch – proceed to or proceeding to an incident. Example: "Salem 4, responding to…" or "Salem 4, respond to…"
Return to, returning to	Used to direct units that are available to a station or other location.

Chapter 4 – Reference

Clear-Text Guide (Continued)

Words and Phrases	Application – Examples
Informational	
Burning operation (specify if illegal)	Indicates a legal fire unless otherwise specified.
Call _____ by phone.	Self-explanatory
Contact _____ message.	Relay message to person named.
Emergency traffic	Used to gain control of the radio frequency to report an emergency in progress or a new incident. Used by Base.
False alarm	Self-explanatory
Fire	Fire emergency requiring a response. Specify structure, field, forest, etc.
Fire under control	Self-explanatory
Is _____ available for a phone call?	Self-explanatory
Let me talk to _____.	Self-explanatory
No smoke or fire	Response to "report on conditions," if appropriate.
Report on conditions	Specify location if needed. Example: "Wise 3 to Lee 2, report on conditions, Jonesville Fire."
Resume normal traffic	Self-explanatory. Used by Base.
Signing on, signing off	Self-explanatory. Used by Base.
Smoke	Suspected or unconfirmed fire.
Weather	Specify report or forecast.
What is your location?	Self-explanatory

ICS Map Display Symbols

From *National Interagency Incident Management System: Basic Land Navigation* (PMS 475, June 2007), chapter 1.

ICS Map Display Symbols (Continued)

Note: GIS symbology from *Geospatial Information Systems Standard Operating Procedures on Incidents*, PMS 936. Current symbology can be found on the NWCG GIS Web site at: http://gis.nwcg.gov/gstop_sop.html.

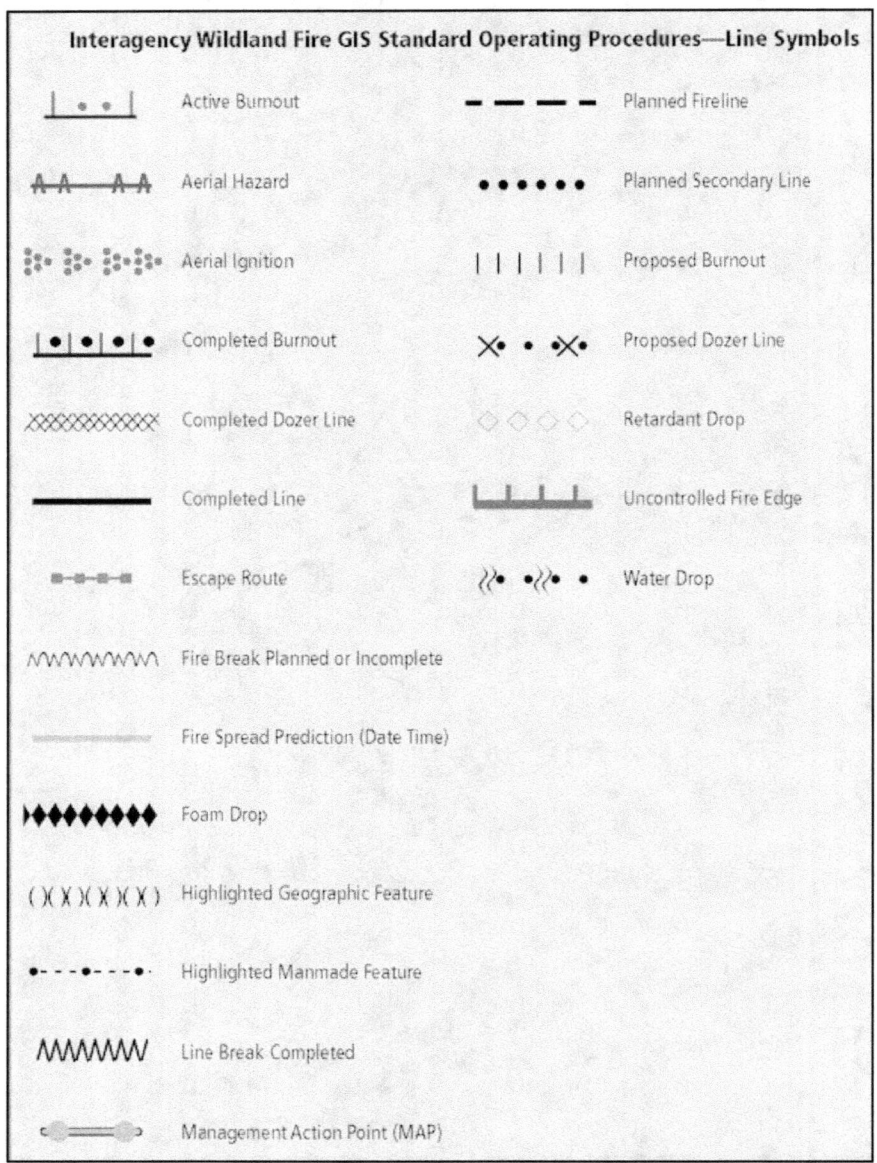

Interagency Wildland Fire GIS Standard Operating Procedures—Polygon-Fill Symbols

- ▨ IR Intense Heat Area
- ▭ IR Heat Perimeter
- ⬚ Maximum Manageable Area (MMA)
- ▭ Temporary Flight Restriction (TFR)

Conversion Factors for Map Scale

Representative Fraction	Inches Per Mile	Inches Per Chain	Feet Per Inch
1: 253,440	1/4	0.00312	21,120
1: 126,720	1/2	0.00625	10,560
1: 63,360	1	0.0125	5,280
1: 31,680	2	0.025	2,640
1: 24,000	2 5/8 or 2.64	0.0328	2,000
1: 21,120	3	0.375	1,760
1: 15,840	4	0.05	1,320
1: 7,920	8	0.10	660

Formula for Area and Circumference of a Circle

Circle, Area	$= 3.1416 \times \dfrac{\text{diameter squared}}{4}$ **or** $= 3.1416 \times \text{radius squared}$
Circle, Circumference	$= 3.1416 \times \text{diameter}$

Acreage Determination Factors

Perimeter Table

Instructions for the use of the following table:

- Use the Perimeter Table as a guide to estimate areas and perimeters. Remember that results are approximate values only and have been rounded off.
- Fires that are roughly circular in shape will have perimeters that approach Minimum values.
- Fires that are very long and narrow or with many fingers will have perimeters that approach or possibly exceed Maximum values.
- Values in the Usual column will represent fires that are oval or wedge shaped.

Acres	Minimum Perimeter (in chains)	Usual Perimeter (in chains)	Maximum Perimeter (in chains)	Acres	Minimum Perimeter (in chains)	Usual Perimeter (in chains)	Maximum Perimeter (in chains)
1	11	17	22	700	300	450	600
2	16	24	32	800	320	475	625
3	19	29	39	900	340	500	675
4	22	34	45	1,000	350	525	700
5	25	38	50	1,200	400	600	775
7	30	45	59	1,400	425	625	850
10	36	53	71	1,600	450	675	900
15	45	65	85	1,800	475	725	950
20	50	75	100	2,000	500	750	1,000
25	55	85	110	2,400	550	825	1,100
30	60	90	125	2,800	600	875	1,175
40	70	105	140	3,200	625	950	1,275
50	80	120	160	3,600	675	1,000	1,350
75	100	150	190	4,000	700	1,075	1,425
100	110	170	220	5,000	800	1,200	1,600
150	140	200	280	6,000	850	1,300	1,700
200	160	240	320	7,000	950	1,400	1,900
300	200	300	400	8,000	1,000	1,500	2,000
400	225	350	450	9,000	1,050	1,600	2,100
500	250	375	500	10,000	1,100	1,700	2,250
600	275	425	550	12,000	1,250	2,000	2,500

Chapter 4 – Reference

Area in Acres

Instructions for the use of the following table:

The Area in Acres table is to help you estimate the area of a fire. To use the table, pace the distance around the fire in chains (1 chain = 66 feet), and determine the general shape of the fire. Select the column (1–6) that best fits the fire's shape, and read the acreage for the paced perimeter shown in the left column.

Explanation of columns representing shapes of fires:

1. Fire in the general shape of a circle.
2. Fire in the shape of either a square or rectangle that is not more than twice as long as it is wide, with a moderately irregular perimeter.
3. Fire in the shape of a rectangle, about three times longer than it is wide. This column also gives the area of a triangle with a moderately irregular perimeter.
4. Fire in the shape of a rectangle about four times longer than it is wide and having a fairly irregular perimeter.
5. Fire that is long and narrow with an irregular perimeter.
6. Fire with two or three long fingers or a very irregular perimeter.

Area in Acres (Continued)

Paced Perimeter	Perimeter in Chains					
	1	2	3	4	5	6
1	.01	.01	.01	.01	.01	.01
2	.03	.02	.02	.02	.01	.01
3	.06	.05	.04	.04	.03	.02
4	.11	.10	.08	.06	.05	.03
5	.17	.15	.12	.10	.07	.05
6	.25	.22	.18	.14	.11	.07
7	.34	.29	.24	.20	.15	.10
8	.45	.38	.32	.26	.19	.13
9	.57	.49	.40	.32	.24	.16
10	.7	.6	.5	.4	.3	.2
12	1.0	.8	.7	.6	.4	.3
14	1.4	1.2	1.0	.8	.6	.4
16	1.8	1.5	1.3	1.0	.8	.5
18	2.3	1.9	1.6	1.3	1.0	.6
20	2.8	2.4	2.0	1.6	1.2	.8
22	3.4	2.9	2.4	1.9	1.4	1.0
24	4.0	3.5	2.9	2.3	1.7	1.2
26	4.7	4.1	3.4	2.7	2.0	1.3
28	5.5	4.7	3.9	3.1	2.3	1.6
30	6.3	5.4	4.5	3.6	2.7	1.8
32	7.2	6.1	5.1	4.1	3.1	2.1
34	8.1	6.9	5.8	4.6	3.5	2.3
36	9.1	7.8	6.5	5.2	3.9	2.6
38	10.1	8.7	7.2	5.8	4.3	2.9
40	11.2	9.6	8.0	6.4	4.8	3.2
42	12.	11.	9.	7.	5.	3.5
44	14.	12.	10.	8.	6.	4.
46	15.	13.	11.	8.5	6.	4.
48	16.	14.	11.5	9.	7.	4.5
50	17.	15.	12.	10.	7.	5.
60	25.	21.	18.	14.	11.	7.
70	34.	30.	25.	20.	15.	10.
80	45.	38.	32.	26.	19.	13.
90	57.	49.	40.	32.	24.	26.
100	70.	60.	50.	40.	30.	20.

Chapter 4 – Reference

Conversion Factors

Linear Measure		
Chain	=	66 feet
	=	100 links
	=	20.1168 meters
Foot (ft)	=	12 inches
	=	0.3048 meters
Inch (in)	=	2.54 centimeters
Kilometer (km)	=	0.62317 statute miles
	=	1,093.6 yards
	=	3,280.8 feet
Link	=	0.66 feet
	=	7.92 inches
	=	0.2012 meters
Meter (m)	=	3.2808 feet
	=	39.37 inches
Mile, statute (mi)	=	5,280 feet
	=	1,760 yards
	=	80 chains
	=	1.60934 kilometers
	=	0.8684 nautical miles
Mile, nautical	=	6,080 feet
	=	2,026.7 yards
	=	92.12 chains
	=	1.8532 kilometers
	=	1.1515 statute miles
Yard (yd)	=	3 feet
	=	36 inches
	=	0.9144 meters

Conversion Factors (Continued)

Square (Area) Measure		
Acre (a)	=	43,560 square feet
	=	4,840 square yards
	=	10 square chains
	=	208.7 x 208.7 feet
	=	0.405 hectares
Hectare (ha)	=	10,000 square meters
	=	2.4 acres
	=	328.1 x 328.1 feet
Square foot (ft^2)	=	144 square inches
Square mile (mi^2)	=	640 acres
Township (T.)	=	36 square miles
	=	6 x 6 miles
Square yard (yd^2)	=	9 square feet
	=	1,296 square inches
Cubic (Volume) Measure		
Cubic foot (ft^3)	=	7.4805 gallons
	=	1,728 cubic inches
	=	28.316 liters
Cubic yard (yd^3)	=	27 cubic feet
	=	200.3 gallons
	=	764.53 liters
Cup	=	8 ounces
Gallon (gal)	=	8.33717 pounds
	=	0.133680 cubic feet
	=	4 quarts
	=	128 ounces
	=	3.7853 liters
Liter (L)	=	0.264179 gallons
	=	1.567 quarts
	=	33.8144 ounces
Pint (pt)	=	2 cups
	=	16 ounces
	=	0.47315 liters
Quart (qt)	=	2 pints
	=	32 ounces
	=	0.9463 liters

Chapter 4 – Reference

Incident Command System Forms

Forms that are routinely used in the Incident Command System (ICS) are listed below. Those marked with an (*) are commonly used in written Incident Action Plans (IAPs).

ICS Form Number	Form Title
201	Incident Briefing
202 (*)	Incident Objectives
203 (*)	Organization Assignment List
204 (*)	Assignment List
205 (*)	Incident Radio Communications Plan
206 (*)	Medical Plan
207	Incident Organizational Chart
209	Incident Status Summary
210	Resource Status Change
211	Incident Check-in List
212	Incident Demobilization Vehicle Safety Inspection
213	General Message
214	Activity Log
215	Operational Planning Worksheet
215A	Incident Action Plan Safety Analysis
216	Radio Requirements Worksheet
217	Radio Frequency Assignment Worksheet
218	Support Vehicle/Equipment Inventory
219	Resource Status Card (T-Card)
220 (*)	Air Operations Summary Worksheet
221	Demobilization Check-out
224	Crew Performance Rating
225	Incident Personnel Performance Rating

Resource Status Card

Colors and Uses

Card Color	Kind of Resource	Form Number
Gray	Headers	219-1
Green	Hand crews	219-2
Rose	Engines	219-3
Blue	Helicopters	219-4
White	Personnel	219-5
Orange	Aircraft, fixed wing	219-6
Yellow	Dozers, tractor plows	219-7
Tan	Miscellaneous equipment and Task Forces	219-8

Distances and Formulas for Estimating Fire Size

Distances

- 1 pace = 2 normal steps
- 11–13 level paces = 1 chain
- 66 feet = 1 chain
- 80 chains = 1 mile
- 10 square chains = 1 acre
- 1 acre = approximately 220 x 220 feet
- 1 acre = 43,560 square feet
- 640 acres = 1 square mile

Formulas

- Area of squares and rectangles = L x W (L = Length, W = Width)
- Area of triangles = ½ (L x W)
- Area of circles = 3.1416 x radius squared
- Compute acres = $\dfrac{\text{Average chains wide x average chains long}}{10 \text{ square chains}}$ = Acres

What the Color and Column of Smoke May Mean

What You See	What It May Mean
The smoke column is thin, rising lazily, and the color is light blue to gray.	Probably a campfire.
The smoke column is narrow, thin, and dark gray to black.	Could be diesel-powered heavy logging or construction equipment.
The smoke column is small, thick, and white in color.	This may mean a small grass fire. If the smoke puffs up every so often, it may mean someone is burning leaves or grass and "feeding" it.
The smoke is widening at the base, and it is predominantly white, but starting to turn brown or black on its downwind side.	This may indicate the fire is spreading in grass and moving unto heavier fuels. Dead brush will burn with a dark brown color; brush with a higher oil content will burn black.
The column of smoke is thick and black, with no spread to the base.	This could be a structure or vehicle fire. It may also be tires burning.
The smoke is black, but some white or light brown is showing away from the main column.	This may mean your vehicle or structure fire has moved into the grass burning.
The column is going straight up.	There is little or no wind on the fire.
The column is going up, but the top of the smoke is bent over.	There is little surface wind, but there is wind where the smoke bends. Beware, that wind may surface at any time.
The smoke is bent over at the ground and building in volume and intensity.	The fire is wind-driven, with a good fuel supply.
The smoke has built to several thousand feet, and a small white cloud has formed on the top.	Fire may be or become plume-dominated, and large fire growth is possible.

Fire Suppression Interpretations from Flame Length

Flame Length	Interpretation
Less than 4 feet	Fires can generally be attacked at the head or flanks by firefighters using hand tools. Handline should hold fire.
4 to 8 feet	Fires are too intense for direct attack on the head with hand tools. Handline cannot be relied on to hold the fire. Bulldozers, engines, and retardant drops can be effective.
8 to 11 feet	Fire may present serious control problems: torching, crowning, and spotting. Control efforts at the head will probably be ineffective.
More than 11 feet	Crowning, spotting, and major fire runs are probable. Control efforts at the head of the fire are ineffective.

Wildland Fire Risk and Complexity Assessment

The Wildland Fire Risk and Complexity Assessment should be used to evaluate firefighter safety issues, assess risk, and identify the appropriate incident management organization. Determining incident complexity is a subjective process based on examining a combination of indicators or factors. An incident's complexity can change over time; incident managers should periodically re-evaluate incident complexity to ensure that the incident is managed properly with the right resources.

Instructions:
Incident Commanders should complete Part A and Part B and relay this information to the Agency Administrator. If the fire exceeds initial attack or will be managed to accomplish resource management objectives, Incident Commanders should also complete Part C and provide the information to the Agency Administrator.

Part A: Firefighter Safety Assessment

Evaluate the following items, mitigate as necessary, and note any concerns, mitigations, or other information.

Evaluate these items	Concerns, mitigations, notes
LCES	
Fire Orders and Watch Out Situations	
Multiple operational periods have occurred without achieving initial objectives	
Incident personnel are overextended mentally and/or physically and are affected by cumulative fatigue.	
Communication is ineffective with tactical resources and/or dispatch.	
Operations are at the limit of span of control.	
Aviation operations are complex and/or aviation oversight is lacking.	
Logistical support for the incident is inadequate or difficult.	

Chapter 4 – Reference

Part B: Relative Risk Assessment

Values				Notes/Mitigation
B1. Infrastructure/Natural/Cultural Concerns **Based on the number and kinds of values to be protected, and the difficulty to protect them, rank this element low, moderate, or high.** Considerations: key resources potentially affected by the fire such as urban interface, structures, critical municipal watershed, commercial timber, developments, recreational facilities, power/pipelines, communication sites, highways, potential for evacuation, unique natural resources, special-designation areas, T&E species habitat, cultural sites, and wilderness.	L	M	H	
B2. Proximity and Threat of Fire to Values **Evaluate the potential threat to values based on their proximity to the fire, and rank this element low, moderate, or high.**	L	M	H	
B3. Social/Economic Concerns **Evaluate the potential impacts of the fire to social and/or economic concerns, and rank this element low, moderate, or high.** Considerations: impacts to social or economic concerns of an individual, business, community or other stakeholder; other fire management jurisdictions; tribal subsistence or gathering of natural resources; air quality regulatory requirements; public tolerance of smoke; and restrictions and/or closures in effect or being considered.	L	M	H	
Hazards				**Notes/Mitigation**
B4. Fuel Conditions **Consider fuel conditions ahead of the fire and rank this element low, moderate, or high.** Evaluate fuel conditions that exhibit high ROS and intensity for your area, such as those caused by invasive species or insect/disease outbreaks; continuity of fuels; low fuel moisture	L	M	H	
B5. Fire Behavior **Evaluate the current fire behavior and rank this element low, moderate, or high.** Considerations: intensity; rates of spread; crowning; profuse or long-range spotting.	L	M	H	
B6. Potential Fire Growth **Evaluate the potential fire growth, and rank this element low, moderate, or high.** Considerations: Potential exists for extreme fire behavior (fuel moisture, continuity, winds, etc.); weather forecast indicating no significant relief or worsening conditions; resistance to control.	L	M	H	
Probability				**Notes/Mitigation**
B7. Time of Season **Evaluate the potential for a long-duration fire and rank this element low, moderate, or high.** Considerations: time remaining until a season ending event.	L	M	H	
B8. Barriers to Fire Spread **If many natural and/or human-made barriers are present and limiting fire spread, rank this element low. If some barriers are present and limiting fire spread, rank this element moderate. If no barriers are present, rank this element high.**	L	M	H	
B9. Seasonal Severity **Evaluate fire danger indices and rank this element low/moderate, high, or very high/extreme.** Considerations: energy release component (ERC); drought status; live and dead fuel moistures; fire danger indices; adjective fire danger rating; preparedness level.	L/M	H	VH/E	
Enter the number of items circled for each column.				

Relative Risk Rating (circle one):

Low	Majority of items are "Low", with a few items rated as "Moderate" and/or "High".
Moderate	Majority of items are "Moderate", with a few items rated as "Low" and/or "High".
High	Majority of items are "High"; A few items may be rated as ""Low" or "Moderate".

Part C: Organization

Relative Risk Rating (From Part B)					
Circle the Relative Risk Rating (from Part B).		L	M	H	
Implementation Difficulty					Notes/Mitigation
C1. Potential Fire Duration Evaluate the estimated length of time that the fire may continue to burn if no action is taken and amount of season remaining. Rank this element low, moderate, or high. Note: This will vary by geographic area.	N/A	L	M	H	
C2. Incident Strategies (Course of Action) Evaluate the level of firefighter and aviation exposure required to successfully meet the current strategy and implement the course of action. Rank this element as low, moderate, or high. Considerations: Availability of resources; likelihood that those resources will be effective; exposure of firefighters; reliance on aircraft to accomplish objectives; trigger points clear and defined.	N/A	L	M	H	
C3. Functional Concerns Evaluate the need to increase organizational structure to adequately and safely manage the incident, and rank this element low (adequate), moderate (some additional support needed), or high (current capability inadequate). Considerations: Incident management functions (logistics, finance, operations, information, planning, safety, and/or specialized personnel/equipment) are inadequate and needed; access to EMS support, heavy commitment of local resources to logistical support; ability of local businesses to sustain logistical support; substantial air operation which is not properly staffed; worked multiple operational periods without achieving initial objectives; incident personnel overextended mentally and/or physically; Incident Action Plans, briefings, etc. missing or poorly prepared; performance of firefighting resources affected by cumulative fatigue; and ineffective communications.	N/A	L	M	H	
Socio/Political Concerns					Notes/Mitigation
C4. Objective Concerns Evaluate the complexity of the incident objectives and rank this element low, moderate, or high. Considerations: clarity; ability of current organization to accomplish; disagreement among cooperators; tactical/operational restrictions; complex objectives involving multiple focuses; objectives influenced by serious accidents or fatalities.	N/A	L	M	H	
C5. External Influences Evaluate the effect external influences will have on how the fire is managed and rank this element low, moderate, or high. Considerations: limited local resources available for initial attack; increasing media involvement, social/print/television media interest; controversial fire policy; threat to safety of visitors from fire and related operations; restrictions and/or closures in effect or being considered; pre-existing controversies/ relationships; smoke management problems; sensitive political concerns/interests.	N/A	L	M	H	
C6. Ownership Concerns Evaluate the effect ownership/jurisdiction will have on how the fire is managed and rank this element low, moderate, or high. Considerations: disagreements over policy, responsibility, and/or management response; fire burning or threatening more than one jurisdiction; potential for unified command; different or conflicting management objectives; potential for claims (damages); disputes over suppression responsibility.	N/A	L	M	H	
Enter the number of items circled for each column.					

Chapter 4 – Reference

Part C: Organization (continued)

Recommended Organization (circle one):

Type 5	Majority of items rated as "N/A"; a few items may be rated in other categories.
Type 4	Majority of items rated as "Low", with some items rated as "N/A", and a few items rated as "Moderate" or "High".
Type 3	Majority of items rated as "Moderate", with a few items rated in other categories.
Type 2	Majority of items rated as "Moderate", with a few items rated as "High".
Type 1	Majority of items rated as "High"; a few items may be rated in other categories.

Rationale:
Use this section to document the incident management organization for the fire. If the incident management organization is different than the Wildland Fire Risk and Complexity Assessment recommends, document why an alternative organization was selected. Use the "Notes/Mitigation" column to address mitigation actions for a specific element, and include these mitigations in the rationale.

Name of Incident:_____ Unit(s):_____

Date/Time:_____ Signature of Preparer:_____

Indicators of Incident Complexity

Common indicators may include the area (location) involved; threat to life, environment and property; political sensitivity, organizational complexity, jurisdictional boundaries, values at risk, and weather. Most indicators are common to all incidents, but some may be unique to a particular type of incident. The following are common contributing indicators for each of the five complexity types.

TYPE 5 INCIDENT COMPLEXITY INDICATORS

General Indicators	Span of Control Indicators
Incident is typically terminated or concluded (objective met) within a short time once resources arrive on scene For incidents managed for resource objectives, minimal staffing/oversight is required One to five single resources may be needed Formal Incident Planning Process not needed Written Incident Action Plan (IAP) not needed Minimal effects to population immediately surrounding the incident Critical Infrastructure, or Key Resources, not adversely affected	Incident Commander (IC) position filled Single resources are directly supervised by the IC Command Staff or General Staff positions not needed to reduce workload or span of control

TYPE 4 INCIDENT COMPLEXITY INDICATORS

General Indicators	Span of Control Indicators
Incident objectives are typically met within one operational period once resources arrive on scene, but resources may remain on scene for multiple operational periods Multiple resources (over 6) may be needed Resources may require limited logistical support Formal Incident Planning Process not needed Written Incident Action Plan (IAP) not needed Limited effects to population surrounding incident Critical Infrastructure or Key Resources may be adversely affected, but mitigation measures are uncomplicated and can be implemented within one Operational Period Elected and appointed governing officials, stakeholder groups, and political organizations require little or no interaction	IC role filled Resources either directly supervised by the IC or supervised through an ICS Leader position Task Forces or Strike Teams may be used to reduce span of control to an acceptable level Command Staff positions may be filled to reduce workload or span of control General Staff position(s) may be filled to reduce workload or span of control

TYPE 3 INCIDENT COMPLEXITY INDICATORS

General Indicators	Span of Control Indicators
Incident typically extends into multiple operational periods Incident objectives usually not met within the first or second operational period Resources may need to remain at scene for multiple operational periods, requiring logistical support Numerous kinds and types of resources may be required Formal Incident Planning Process is initiated and followed Written Incident Action Plan (IAP) needed for each Operational Period Responders may range up to 200 total personnel Incident may require an Incident Base to provide support Population surrounding incident affected Critical Infrastructure or Key Resources may be adversely affected and actions to mitigate effects may extend into multiple Operational Periods Elected and appointed governing officials, stakeholder groups, and political organizations require some level of interaction	IC role filled Numerous resources supervised indirectly through the establishment and expansion of the Operations Section and its subordinate positions Division Supervisors, Group Supervisors, Task Forces, and Strike Teams used to reduce span of control to an acceptable level Command Staff positions filled to reduce workload or span of control General Staff position(s) filled to reduce workload or span of control ICS functional units may need to be filled to reduce workload

TYPE 2 INCIDENT COMPLEXITY INDICATORS

General Indicators	Span of Control Indicators
Incident displays moderate resistance to stabilization or mitigation and will extend into multiple operational periods covering several days Incident objectives usually not met within the first several Operational Periods Resources may need to remain at scene for up to 7 days and require complete logistical support Numerous kinds and types of resources may be required including many that will trigger a formal demobilization process Formal Incident Planning Process is initiated and followed Written Incident Action Plan (IAP) needed for each Operational Period Responders may range from 200 to 500 total Incident requires an Incident Base and several other ICS facilities to provide support Population surrounding general incident area affected Critical Infrastructure or Key Resources may be adversely affected, or possibly destroyed, and actions to mitigate effects may extend into multiple Operational Periods and require considerable coordination Elected and appointed governing officials, stakeholder groups, and political organizations require a moderate level of interaction	IC role filled Large numbers of resources supervised indirectly through the expansion of the Operations Section and its subordinate positions Branch Director position(s) may be filled for organizational or span of control purposes Division Supervisors, Group Supervisors, Task Forces, and Strike Teams used to reduce span of control All Command Staff positions filled All General Staff positions filled Most ICS functional units filled to reduce workload

TYPE 1 INCIDENT COMPLEXITY INDICATORS

General Indicators	Span of Control Indicators
Incident displays high resistance to stabilization or mitigation and will extend into numerous operational periods covering several days to several weeks Incident objectives usually not met within the first several Operational Periods Resources may need to remain at scene for up to 14 days, require complete logistical support, and several possible personnel replacements Numerous kinds and types of resources may be required, including many that will trigger a formal demobilization process DOD assets, or other nontraditional agencies, may be involved in the response, requiring close coordination and support Complex aviation operations involving multiple aircraft may be involved Formal Incident Planning Process is initiated and followed. Written Incident Action Plan (IAP) needed for each Operational Period Responders may range from 500 to several thousand total Incident requires an Incident Base and numerous other ICS facilities to provide support Population surrounding the region or state where the incident occurred is affected Numerous Critical Infrastructure or Key Resources adversely affected or destroyed. Actions to mitigate effects will extend into multiple Operational Periods spanning days or weeks and require long-term planning and considerable coordination Elected and appointed governing officials, stakeholder groups, and political organizations require a high level of interaction	IC role filled Large numbers of resources supervised indirectly through the expansion of the Operations Section and its subordinate positions Branch Director Position(s) may be filled for organizational or span of control purposes Division Supervisors, Group Supervisors, Task Forces, and Strike Teams used to reduce span of control All Command Staff positions filled and many include assistants All General Staff positions filled and many include deputy positions Most or all ICS functional units filled to reduce workload

ACRONYMS

Acronyms used in this document are as follows:

Acronym	Meaning
AAR	After Action Review
ABRO	Aircraft Base Radio Operator
ADP	Automatic Data Processing
AGL	above ground level
AOBD	Air Operations Branch Director
AREP	Agency Representative
ASGS	Air Support Group Supervisor
ASM	Aerial Supervision Module
ATCO	Air Tanker/Fixed Wing Coordinator
ATGS	Air Tactical Group Supervisor
BCMG	Base/Camp Manager
CAF	compressed air foam
CAFS	compressed air foam system
CDL	Commercial Driver's License
CLMS	Claims Specialist
CMSY	Commissary Manager
CO	carbon monoxide
COML	Communications Unit Leader
COMP	Compensation/Claims Unit Leader
COMT	Incident Communications Technician
COR	Contracting Officer Representative
COST	Cost Unit Leader
COTR	Contracting Officer's Technical Representative
CREP	Crew Representative
CRWB	Crew Boss, Single Resource
CWPP	Community Wildfire Protection Plan
DA	Density Altitude
DECK	Deck Coordinator

Acronym	Meaning
DIVS	Division/Group Supervisor
DMOB	Demobilization Unit Leader
DOCL	Documentation Unit Leader
DOI	Department of the Interior
DOT	Department of Transportation
DPRO	Display Processor
ENGB	Engine Boss, Single Resource
EQPM	Equipment Manager
EQTR	Equipment Time Recorder
ERFOG	Emergency Responder Field Operating Guide
ETA	estimated time of arrival
ETD	estimated time of departure
FAA	Federal Aviation Administration
FACL	Facilities Unit Leader
FARSITE	Fire Area Simulator
FBAN	Fire Behavior Analyst
FDUL	Food Unit Leader
FELB	Felling Boss, Single Resource
FEMA	Federal Emergency Management Agency
FEMO	Fire Effects Monitor
FIRB	Firing Boss, Single Resource
FFT1	Firefighter Type 1
FFT2	Firefighter Type 2
FL	Friction Loss
FOBS	Field Observer
FSC1/2	Finance/Administration Section Chief Type 1/2
FSPro	Fire Spread Probability
FTA	Fire Traffic Area
GACC	Geographic Area Coordination Center
GIS	Geographic Information System
GISS	Geographic Information System Specialist

Acronym	Meaning
GSUL	Ground Support Unit Leader
GVWR	gross vehicle weight rating
H	Head
HazMat	Hazardous material
HEB1	Helibase Manager Type 1
HEB2	Helibase Manager Type 2
HECM	Helicopter Crewmember
HEQB	Heavy Equipment Boss, Single Resource
HESM	Helispot Manager
HLCO	Helicopter Coordinator
HMGB	Helicopter Manager, Single Resource
HP	horsepower
HRO	high reliability organization
HRSP	Human Resource Specialist
IAP	Incident Action Plan
IARR	Interagency Resource Representative
IBA1/2	Incident Business Advisor Type 1/2
IC	Incident Commander
ICC	Incident Communications Center
ICP	Incident Command Post
ICPI	Incident Contract Project Inspector
ICS	Incident Command System
ICT1/2/3/4/5	Incident Commander
ICT4	Initial Attack Incident Commander Type 4
ICT5	Initial Attack Incident Commander Type 5
IHC	Interagency Hotshot Crew
IMET	Incident Meteorologist
IMT	Incident Management Team
INCM	Incident Communications Center Manager
INJR	Compensation-For-Injury Specialist
IP	Initial point

Acronym	Meaning
IRIN	Infrared Interpreter
IRPG	Incident Response Pocket Guide
LCES	Lookout(s), Communication(s), Escape(s), and Safety Zone(s)
LOAD	Loadmaster
LOFR	Liaison Officer
LPG	liquefied petroleum gas
LSC1/2	Logistics Section Chief
LTAN	Long Term Fire Analyst
MAFFS	Modular Airborne Fire Fighting System
MAV	Minimum Acceptable Vehicle
MEDL	Medical Unit Leader
NFPA	National Fire Protection Association
NIMS	National Incident Management System
NP	Nozzle Pressure
NWCG	National Wildfire Coordinating Group
OPBD	Operations Branch Director
ORDM	Ordering Manager
OSHA	Occupational Safety and Health Administration
OWDC	Operations and Workforce Development Committee
PARK	Parking Tender
PDP	Pump Discharge Pressure
PIO1/2	Public Information Officer
PIOF	Public Information Officer
PPE	personal protective equipment
PROC	Procurement Unit Leader
PSC1/2	Planning Section Chief
psi	pounds per square inch
PTRC	Personnel Time Recorder
R&R	rest and recuperation
RADO	Radio Operator
RCDM	Receiving/Distribution Manager

Acronym	Meaning
RERAP	Rare Event Risk Assessment Process
RESL	Resources Unit Leader
RH	relative humidity
RTI	radio telephone interconnect
SAFECOM	Aviation Safety Communiqué
SCKN	Status/Check-in Recorder
SEAT	Single Engine Air Tanker
SECM	Security Manager
SEMG	Single Engine Air Tanker Manager
SITL	Situation Unit Leader
SOFR	Safety Officer
SOPL	Strategic Operational Planner
SPUL	Supply Unit Leader
STAM	Staging Area Manager
STPS	Structure Protection Specialist
STCR	Strike Team Leader Crew
STEN	Strike Team Leader Engine
STEQ	Strike Team Leader Heavy Equipment
SUBD	Support Branch Director
SVBD	Service Branch Director
T&E	threatened and endangered
TBSA	Total Body Surface Area
TFLD	Task Force Leader
TFR	Temporary Flight Restriction
TIME	Time Unit Leader
TNSP	Incident Training Specialist
TOLC	Takeoff and Landing Coordinator
UN	United Nations
USFS	U.S. Forest Service
VHF-AM	Very High Frequency-Amplitude Modulation
VHF-FM	Very High Frequency-Frequency Modulation

Acronym	Meaning
VIPs	very important persons
WFDSS	Wildland Fire Decision Support System